Immobilized Biomolecules in Analysis

The Practical Approach Series

SERIES EDITOR

B. D. HAMES
Department of Biochemistry and Molecular Biology
University of Leeds, Leeds LS2 9JT, UK

See also the Practical Approach web site at **http://www.oup.co.uk/PAS**

★ **indicates new and forthcoming titles**

Immobilized Biomolecules in Analysis

A Practical Approach

Edited by

TONY CASS

Reader in Bioanalytical Chemistry at
Imperial College, London

and

FRANCES S. LIGLER

Head of the Biosensors and Biomaterials Laboratory,
Naval Research Laboratory, Washington

OXFORD
UNIVERSITY PRESS

OXFORD

UNIVERSITY PRESS

Great Clarendon Street, Oxford OX2 6DP

Oxford University Press is a department of the University of Oxford
and furthers the University's aim of excellence in research, scholarship,
and education by publishing worldwide in

Oxford New York

Athens Auckland Bangkok Bogotá Buenos Aires Calcutta
Cape Town Chennai Dar es Salaam Delhi Florence Hong Kong Istanbul
Karachi Kuala Lumpur Madrid Melbourne Mexico City Mumbai
Nairobi Paris São Paulo Singapore Taipei Tokyo Toronto Warsaw

and associated companies in Berlin Ibadan

Oxford is a registered trade mark of Oxford University Press

Published in the United States
by Oxford University Press Inc., New York

© Oxford University Press, 1998

Users of books in the Practical Approach Series are advised that prudent
laboratory safety procedures should be followed at all times. Oxford
University Press makes no representation, express or implied, in respect of
the accuracy of the material set forth in books in this series and cannot
accept any legal responsibility or liability for any errors or omissions
that may be made.

A catalogue record for this book is available from the British Library

Library of Congress Cataloging in Publication Data
(Data available)

ISBN 0 19 963637 0 (Hbk)
ISBN 0 19 963636 2 (Pbk)

Typeset by Footnote Graphics, Warminster, Wilts
Printed in Great Britain by Information Press, Ltd, Eynsham, Oxon.

Preface

Biosensors herald the coming of a technology that will explode during the next decade; they demonstrate that we can harness the incredible functions of living molecules and cells, crafted for millennia by nature, for our own, more limited purposes. Nevertheless, to make use of these small wonders, we have to first capture them and restrict them geographically to the artificial space in which we demand that they perform. Moreover, we have to do this without jeopardizing the ability of the relatively fragile molecule or cell to do the desired job.

Over the last decade, we have both attended numerous conferences labelled 'Biosensors' in which over 50% of the papers centred on how to immobilize a biomolecule or cell, i.e. a biomaterial, on a sensing surface. In many cases, the papers presented documented the loss of activity suffered by the biomaterial when subjected to glutaraldehyde or other abusive chemical treatment. In other instances, the loss of activity was simply a function of non-specific interactions of the biomaterial with the surface or steric hindrances placed on the molecule by the method of immobilization. There have been discussions *ad nauseam* about how to immobilize a biomolecule in the 'right orientation', but the subsequent work often seemed to be more driven by cartoons that could be drawn on a computer than by any deep appreciation of how the biomolecule or its function was being affected by the surface to which is was attached.

And what is really happening to these harnessed biomolecules? The portentous words of Enrico Fermi have become a favourite quote for both of us, 'God made the solid state. He left the surface to the Devil'. Consider the poor biomolecule or cell trying to do its job in its new location. It has to cope with a foreign surface, which often—like a hot skillet fries an egg—tries to alter its conformation mainly through hydrophobic interactions and so denature it. It may have to cope with tethers that—like those the Lilliputians used to immobilize Gulliver—tie it down via many small lines. It has to function in a geometrically deranged configuration where solution mass transfer rules usually do not apply. A variety of solutes are present which may affect not only the immobilized cell or molecule but also the analytes with which it is supposed to interact. And it is probably being asked to perform lying down or standing on its head!

Both of us, frustrated with the lack of any recently compiled information on biomolecule immobilization, have been considering such a book as this for several years. Hopefully, it will become obsolete in the next decade as we deepen our understanding of the factors controlling biomolecular interactions with surfaces and biomolecular function at surfaces. However, we have tried to provide the reader with a set of options to guide his choice of how to immobilize a biomolecule so that it can best perform the desired function. Included are chapters on biomolecule immobilization via adsorption (Johnston and Ratner), entrapment (Dave *et al.*), through a tried-and-true tether (Shriver-Lake), and via site-specific binding (Egodage and Wilson), or avidin–biotin technology (Wilchek and Bayer). Elegant surface chemistries such as self-assembled monolayers (Liedberg and Cooper) or conducting organic polymers (Schuhmann) can be recruited. An exciting new approach describes a method for immobilization that reversibly regulates function (Stayton and Hoffman). Bright describes a method for monitoring the biomolecule once it is immobilized, and Karlsson and Löfås describe how kinetic analysis is important in understanding the behaviour of these systems. The advances in cell patterning which require both spatially and chemically controlled immobilization are described by Clark.

We expect the field of immobilized biomolecules and cells to expand far beyond biosensor applications. Already there are examples of immobilized biomolecules used for pharmaceutical production and processing. Over the coming decades, 'biomaterials' will become a term not limited to the scientific community as immobilized biomolecules and cells are incorporated into 'smart polymers' which can respond to their environment, into filters for remediating toxic pollutants, into artificial tissues for organ replacement, and into silicon chips for advanced computing. Immobilizing the biomolecule in a functional state is the critical starting point.

London   T.C.
Washington, DC   F.S.L.
May 1998

Contents

Contents

Contents

Contents

Contributors

EDWARD A. BAYER
Department of Biological Chemistry, The Weizmann Institute of Science, Rehovot 76100, Israel.

FRANK V. BRIGHT
Department of Chemistry, Natural Sciences Complex, State University of New York at Buffalo, Buffalo, NY 14260–3000, USA.

PETER CLARK
Biomedical Sciences, Imperial College School of Medicine, South Kensington, London SW7 2AZ, UK.

JONATHAN M. COOPER
Department of Electronics and Electrical Engineering, University of Glasgow, Glasgow, Scotland, UK.

BAKUL C. DAVE
Department of Chemistry and Biochemistry, Southern Illinois University at Carbondale, Carbondale, IL 62901–4409, USA.

BRUCE DUNN
Department of Materials Science and Engineering, University of California, Los Angeles, CA 90095, USA.

KAMAL L. EGODAGE
HBC-CT Drug Delivery, 115 McCollum Research Laboratories, University of Kansas, Lawrence, KS 66045, USA.

ALLAN S. HOFFMAN
University of Washington, Center for Bioengineering, Box 351750, Seattle, WA 98195, USA.

ERIKA JOHNSTON
Center for Biomolecular Science & Engineering, Code 6910, Naval Research Laboratory, Washington, DC 20375–5348, USA.

R. KARLSSON
Biacore AB, Rapsgatan 7, S-754 50 Uppsala, Sweden.

BO LIEDBERG
Molecular Films & Surface Analysis Group, Laboratory of Applied Physics, Linköping University, Sweden, S-58183.

S. LÖFÅS
Biacore AB, Rapsgatan 7, S-754 50 Uppsala, Sweden.

Contributors

BUDDY D. RATNER
University of Washington Engineered Biomaterials, University of Washington, Center for Bioengineering, Box 351750, Seattle, WA 98195, USA.

WOLFGANG SCHUHMANN
Fakultät für Chemie, Lehrstuhl für Analytische Chemie, Ruhr-Universität Bochum, D-44780 Bochum, Germany.

LISA C. SHRIVER-LAKE
Center for Bio/Molecular Science and Engineering, Naval Research Laboratory, Washington, DC 20375–5348, USA.

PATRICK S. STAYTON
Molecular Engineering Program, University of Washington, Center for Bioengineering, Box 351750, Seattle, WA 98195, USA.

JOAN S. VALENTINE
Department of Chemistry and Biochemistry, UCLA, Los Angeles, CA 90095, USA.

MEIR WILCHEK
Department of Biological Chemistry, The Weizmann Institute of Science, Rehovot 76100, Israel.

GEORGE S. WILSON
Department of Chemistry, 2010 Malott Hall, University of Kansas, Lawrence, KS 66045, USA.

JEFFREY I. ZINK
Department of Chemistry and Biochemistry, UCLA, Los Angeles, CA 90095, USA.

Abbreviations

AAc	acrylic acid
Ab	antibody
Ac	acrylodan (6-acryloyl-(dimethylamino)-naphthalene)
AIBN	2,2′-azoisobutyronitrile
AOT	dioctyl sodium sulfosuccinate
APTS	aminopropyltriethoxysilane
BCHZ	biocytin hydrazide
BHZ	biotin hydrazide
BMA	butyl methacrylate
BNHS	biotinyl *N*-hydroxysuccinimide
BPF	bandpass filter
BSA	bovine serum albumin
B-sulfo-NHS	biotinyl N-hydroxy-sulfo-succinimide ester
BxHZ	biotinyl ε-aminocaproyl hydrazide
BxNHS	biotinyl ε-aminocaproyl *N*-hydroxysuccinimide
Bx-sulfo-NHS	biotinyl ε-aminocaproyl *N*-hydroxy-sulfo-succinimide ester
CDI	carbonyldiimidazole
CMEC	1-cyclohexyl-3-(2-morpholinoethyl)carboxdiimide
CNBr	cyanogen bromide
DBB	dibenzoyl biocytin
DMF	dimethylformamide
DMSO	dimethylsulfoxide
dNTP	deoxynucleotide triphosphate
DO	dissolved oxygen
DTT	dithiothreitol
ECM	extracellular matrix
EDC	1-ethyl-3-(3-dimethylaminopropyl)carbodiimide
EDTA	ethylenediaminetetraacetic acid
ELISA	enzyme-linked immunosorbent assay
EO	ethylene oxide
FITC	fluorescein isothiocyanate
FMP	2-fluoro-1-methylpyridinium toluene-4-sulfonate
GMBS	gamma-maleimidobutrylsuccinimide
GOPS	glycidoxypropyltrimethoxysilane
GOx	glucose oxidase
HBS	hepes-buffered saline
HRP	horseradish peroxidase
HSA	human serum albumin
iPA	isopropyl alcohol
LDPE	low density polythylene

M$_2$C$_2$H	4-(*N*-maleimidomethyl)-cyclohexane-1-carboxylhydrazide–HCl
MAb	monoclonal antibody
MBDD	12-mercapto (8-biotinamide-3,6-dioxaocytl)dodecanamide
Mb	myoglobin
MbCO	carbonyl myoglobin
MbO$_2$	*oxy*myoglobin
2-MEA	2-mercaptoethylamine
MeOH	methanol
MPB	maleimidopropionyl biocytin
MPC	2-methacryloyl oxyethyl phosphoryl choline
MPM	multifrequency phase modulation
MPTS	mercaptopropyltrimethoxysilane
NAS	*N*-acryloxysuccinimide
NHS	*N*-hydroxysuccinimide ester
NIPAAm	*N*-isopropylacrylamide
OF	optical fibre
OTS	octadecyl trichlorosilane
PBS	phosphate-buffered saline
PCR	polymerase chain reaction
4-PDS	4,4′-dithiodipyridine
PEO	poly(ethylene oxide)
PET	poly(ethylene terephthalate)
PMMA	poly(methyl methacrylate)
PNIPAAm	poly(*N*-isopropylacrylamide)
poly HEMA	poly (hydroxyethylmethacrylate)
PU	polyurethane
RF	radiofrequency
RU	resonance units
SAM	self-assembled monolayer
SAW	surface acoustic wave
SPIN	surface physical interpenetrating network
SPR	surface plasmon resonance
TBATos	tetrabutylammonium toluene-4-sulfonate
TCEP	Tris(2-carboxyethyl)phosphine
TFAA	trifluoroacetic acid
THF	tetrahydrofuran
TIRF	total internal reflection fluorescence
TMOS	tetramethyl orthosilicate
TNBS	2,4,6-trinitrobenzenesulfonic acid
TOF	time-of-flight
UHV	ultrahigh vacuum
VS	vinyl sulfone
VTPDMS	vinyl terminated polydimethylsiloxane
XPS	X-ray photoelectron spectroscopy

1

Silane-modified surfaces for biomaterial immobilization

1. Introduction

Biosensors, affinity chromatography, and many bioanalytical methods require a high density of functional molecules, low non-specific protein adsorption, long-term stability, and durability (1). Proteins, including antibodies and enzymes, and cells have been immobilized onto solid supports for these and other applications over the last 30 years. In the biosensor arena, optical transduction methods are increasingly employed. With this increase, solid supports such as fused silica and quartz are appropriate substrates for immobilization. Non-specific protein adsorption to these substrates has been a major problem.

Adsorption, entrapment, and covalent attachment are the leading techniques employed for immobilization of biomolecules onto solid supports. Adsorption of biomolecules is simple and mild. This ease of attachment is somewhat offset by the denaturation of biomolecules over time as well as leaching off the surface. Ulbrich *et al.* found that adsorption does not always provide as high a density of protein on the surface, as does covalent attachment (2). Entrapment of biomolecules in a polymer or within a membrane-sealed bag appears to be straightforward, but in fact, there are technical hurdles to overcome in maintaining the integrity of the system. If the analyte is large, movement through the polymer or membrane may be restricted. Covalent attachment is the preferred method for biomolecule binding. High surface density of the immobilized biomolecules has been achieved. In addition to a reduction in protein leaching, it has been suggested that covalent binding under certain conditions can improve apparent protein stability (3, 4). Primary amines, thiols, and carbohydrate groups of the biomolecules are often the sites of immobilization.

Factors to consider when choosing an immobilization method are the effect on the orientation of the biomaterial, particularly the active site. Organo-functional silanes have been employed as coupling agents to attach organic compounds (proteins) to inorganic substrates (silica, quartz, platinum, etc.) (5). There are many ways to silanize surfaces and covalently attach biomolecules. This paper will concentrate on providing protocols for some of the major methodologies currently in use. This paper is not to be considered a review of all methods for immobilization of biomaterials onto silane-modified surfaces.

2. General concepts

2.1 Silanization

For many of the sensor applications, fused silica, quartz, or glass are used as the solid supports. A problem of non-specific protein adsorption exists for these surfaces. Silanes have been used to modify surfaces both to reduce non-specific adsorption and to provide moieties suitable for covalent attachment. Many of the solid supports have or can be modified to contain surface hydroxyls which react with methoxy/ethoxy residues of the silanes. The other end of the silane provides a reactive residue to bind to biomolecules or cross-linkers.

There are three major silane compounds employed for the immobilization of biomaterials. The first one, mercaptopropyltrimethoxysilane, has a terminal thiol group. The second silane is aminopropyltriethoxysilane which has a terminal amine group. The last major silane is glycidoxypropyltrimethoxysilane which contains an epoxy residue for further reactions. Factors in determining which silane to employ include choice of reactive group on the biomolecule (amine, thiol, carbohydrate) and type of cross-linker (homobifunctional or heterobifunctional).

There are two primary procedures for silanization of solid supports. One method is immersion into a silane solution and is referred to as a liquid phase method. The other procedure is vapour deposition of the silane onto the solid support. This article will concentrate on the immersion or liquid phase method. With this procedure, the solid support is placed into a 2–10% silane solution in water or organic solvent. This method lends itself for use by a variety of scientists. The vapour deposition method, on the other hand, requires refluxing an organic solvent containing the silane to generate a silane vapour. This procedure is easier to perform for someone with experience in organic chemistry. It should be noted that silanes, if not properly stored, do degrade. These degraded silanes will not provide a uniform monolayer on the solid support.

2.2 Solid support surfaces

No matter which silanization method is employed, cleaning of the solid support to generate reactive hydroxyl groups is critical for effective immobilization of biomaterials. There are several types of Si–OH groups that can form on silica surfaces. The geminal and isolated silanols are reactive, whereas the vicinal silanol and the siloxane groups are not (*Figure 1*). If the surface is not properly cleaned of oils, dirt, detergents, etc., the reactive hydroxyl groups will not be formed, and the silane will not be deposited in a uniform manner.

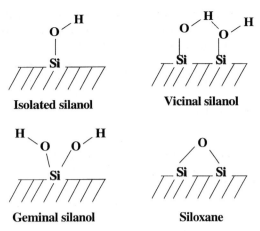

Figure 1. Types of hydroxyl group on silica surfaces. These are four possible types of SiOH groups on silica surfaces in which only the geminal and isolated silanols are reactive.

Many comments against silanization methods concern the non-uniformity of the silane monolayer. The major contributors to such non-uniformity are inadequate surface cleaning and decomposure or polymerization of the silane.

2.3 Cross-linkers

Over 300 cross-linkers are currently available for protein conjugation (6). There are two groups of cross-linkers, homobifunctional and heterobifunctional. A cross-linker is a molecule which has two reactive groups with which to covalently attach proteins or other molecules. In between the reactive groups is the 'bridge' or spacer group. Steric interference with the activity of the biomolecule by the surface may be ameliorated by altering the bridge composition. To reduce intramolecular cross-linking, it is advisable to used a larger/longer bridge. In the case of heterobifunctional cross-linkers, the reactive groups have two dissimilar functionalities of different specificities. With heterobifunctional cross-linkers, linking can be controlled selectively and sequentially since it is performed in two steps. On the other hand, the homo-bifunctional cross-linker's reactive groups are the same. This type of cross-linking is usually performed as a one-step procedure. Chances of unwanted inter- and intramolecular cross-linking exist with the homobifunctional cross-linkers. The immobilization procedure must provide a means for controlling some of the unwanted cross-linking. A thorough review of cross-linking can be found in an article by Mattson *et al.* (7) and the book by Wong (6).

2.4 Biomolecules

Many types of biomolecules have been employed in biosensors, affinity chromatography techniques, and other surface analysis procedures. Enzymes and antibodies are considered the primary biomolecules used in these areas. Whole cells and receptors have also been used. These biomolecules contain a number of reactive side chains by which they can be immobilized onto solid supports. The most common reactive groups in proteins are amines from lysines, and α-amino groups and thiols from cystine, cysteine, and methionine (8). The thiols are very reactive and are often found as bridged disulfide groups. To be able to use the thiols of proteins, the disulfide bond must be reduced with a reagent such as dithiothreitol (DTT). This reduction can also have some unwanted effects on the activity of the protein.

In some proteins, a carbohydrate residue may also be present. In the case of antibodies, this carbohydrate group is located away from the active site. The carbohydrate can be oxidized to an aldehyde with reagents such as sodium meta-periodate. The aldehyde can then be reacted with hydrazide containing cross-linkers for immobilization. Again, modification of the protein prior to immobilization can cause reduced activity. For antibodies, immobilization through the carbohydrate group can prevent binding at the active site and provide some orientation to the immobilized biomolecule.

3. Methods for preparing surfaces for silanization

As mentioned above, a critical part of covalently attaching a biomolecule to the surface is the preparation of the solid support to generate reactive hydroxyls. There are many protocols available, varying from very complex to simple procedures. It should be noted that the surfaces should be cleaned just prior to silanization for optimum immobilization. Following are a sampling of cleaning procedures. The first cleaning protocol (*Protocol 1*) described by Bhatia *et al.* (9) is a simple procedure involving acids and methanol at room temperature and boiling water. This procedure can be accomplished in less than two hours and performed in all types of laboratories; preferably those with a chemical hood to remove acid fumes. As with all procedures employing acids, safe utilization includes the wearing of appropriate gloves and protective eyewear.

Protocol 1.   Substrate cleaning with hydrochloric acid, methanol, and sulfuric acid

Reagents

- Hydrochloric acid (HCl)
- Methanol (MeOH)

- Sulfuric acid, concentrated (H_2SO_4)
- Lint-free clean room paper

Method

1. Immerse the substrate into a mixture of HCl:MeOH (1:1, v/v) at room temperature for 30 min.

---

2. Rinse the substrate three times with distilled water.

3. Immerse the substrate into concentrated H_2SO_4 at room temperature for 30 min.

4. Rinse the substrate well three times with distilled water.[a]

5. Boil the substrate for 30 min in boiling water.

6. Dry the substrate on lint-free clean room paper, or air dry.

[a] It is important to remove all H_2SO_4 residue from the substrate in order to make a uniform silane monolayer.

Several cleaning procedures employ ammonia hydroxide and hydrogen peroxide at elevated temperatures as in *Protocol 2* (10). These protocols make the surface hydrophilic and ready for silanization. Safety precautions should be taken when working with a hot basic solution to protect the worker from fumes and skin burns. Hong *et al.* (11) used a modified version of *Protocol 2*. The changes include ammonia hydroxide solution and the use of concentrated sulfuric acid instead of hydrochloric acid.

Protocol 2. Substrate cleaning with ammonia hydroxide, hydrogen peroxide, and hydrochloric acid

Reagents
- Ammonia hydroxide (NH_4OH)
- Hydrogen peroxide (30%) (H_2O_2)
- Hydrochloric acid (HCl)
- Nitrogen gas

Method

1. Immerse substrate into a solution of $NH_4OH:H_2O_2:H_2O$ (1:1:5) at 80°C for 5 min.

2. Rinse the substrate with distilled water.

3. Immerse the substrate into a solution of $HCl:H_2O_2:H_2O$ (1:1:5) at 80°C for 5 min.

4. Rinse with distilled water and dry by passing nitrogen gas over the surface.

The last cleaning method (*Protocol 3*) uses hot chromic acid and requires an oven to make the surface hydrophilic (12, 13). A concern exists with using hot chromic acid. Safety precautions, due to the hot acid, need to be observed with this type of procedure. It has been suggested that residues are left on the solid support after chromic acid cleaning. Rinsing thoroughly and not touching the surface are important to prevent residues after cleaning.

Protocol 3.   Substrate cleaning with chromic acid

Equipment and reagents
- Oven/desiccator
- Chromic acid

Method

1. Immerse the solid support into hot chromic acid at 80°C for 30 min.

2. Rinse the substrate thoroughly with purified deionized water.[a]

3. Dry the solid support in a oven/desiccator at 120°C for more than 2 h.

[a] The substrate should not be touched during this and the next step to maintain the clean surface.

---

4. Immobilization procedures using mercapto-terminal silanes

Most biomolecules contain primary amines to which cross-linkers can be attached without prior modification of the biomolecule. This is important to maintaining antibody and enzyme activity. Modification of biomolecules results in lower activity and loss of protein during the modification procedure. By using a mercapto-terminal silane on the surface, heterobifunctional cross-linkers can be employed which react with both the thiols on the surface and the amines on the biomolecules. In 1989, Bhatia *et al.* reported a method for this type of immobilization (9, 14). Several other groups have reported using this method with high densities of active protein attached to the surface (15–17).

It is important with this protocol that it be done at one time. Exposure of a mercapto-silanized solid support to air causes a slow oxidation of the thiols. After being exposed to air overnight, up to 50% of the thiols are no longer available to react with the cross-linker. Antibody coated glass slides have been stored in phosphate-buffered saline at room temperature for up to one year with less than 20% loss of activity (4). The preferred method for storage of the biomolecule immobilized support is refrigeration, but this is not critical. This protocol demonstrates the ability of covalent immobilization to improve stability of the antibody compared to that in solution.

Other heterobifunctional cross-linkers (Pierce Chemical Co.) have been used in this protocol with similar results (9, 18). Antibody densities of 1.0–2.2 ng/mm^2 and 25–35% antibody activity have been achieved using this protocol.

Proteins have been immobilized in patterns employing this chemistry (19). After silanization with the mercapto-terminal silane, the area when no protein is desired is irradiated with 193 nm light. This light converts the thiols to sulfonates which resist protein absorption. In the areas unirradiated, proteins were coupled using *Protocol 4*, steps 3–6.

Protocol 4. Protein immobilization with mercaptopropyltrimethoxysilane and heterobifunctional cross-linkers

Reagents

- Mercaptopropyltrimethoxysilane (MPTS) (Fluka)
- Toluene
- Ethanol
- Dimethylformamide (DMF)
- Phosphate-buffered saline pH 7.4 (PBS)
- γ-Maleimidobutyryloxysuccinimide (GMBS): dissolve cross-linker in DMF and adjust its concentration to 2 mM with ethanol (Fluka)
- Protein: 0.05 mg/ml in PBS
- Nitrogen gas

Method

1. Immerse the cleaned substrate into a 2% MPTS solution in toluene under a nitrogen atmosphere for 0.5–2 h. This can be accomplished in a glove bag.

2. Rinse the silanized substrate in toluene and air dry.

3. Immerse the silanized substrate into a 2 mM GMBS solution for 1 h at room temperature in a closed container to prevent evaporation.

4. Rinse the substrate with PBS thoroughly three times.

5. Immerse the substrate into a 0.05 mg/ml protein solution in PBS for 1 h at room temperature.

6. Rinse the protein coated substrate in PBS and store.

A modification of *Protocol 4* for use with carbohydrate reactive hetero-bifunctional cross-linkers has been tested with similar densities and activities (*Protocol 5*) (18). By using carbohydrate reactive cross-linkers, the active site of the antibody is not employed for immobilization. The protein is modified by changing the carbohydrate into an aldehyde, prior to reaction with the hydrazide residue of the cross-linker. As mentioned earlier, modification to biomolecules can result in reduced activity and loss of the biomolecule.

Protocol 5. Protein immobilization with mercaptopropyltrimethoxysilane and carbohydrate reactive heterobifunctional cross-linkers

Equipment and reagents

- Microfilterfuge tubes (Rainin Instrument Co. Inc.)
- Microcentrifuge, fixed angle
- Shaker
- Mercaptopropyltrimethoxysilane (MPTS) (Fluka)

Protocol 5.   *Continued*

- Toluene
- Sodium meta-periodate (Sigma Chemical Corp.)
- Glycerol (Sigma)
- Dimethyl sulfoxide (DMSO)
- Reaction buffer: 0.1 M sodium acetate pH 5.5

- Carbohydrate reactive heterobifunctional cross-linker: 4-(N-maleimidomethyl)-cyclo-hexane-1-carboxylhydrazide–HCl (M$_2$C$_2$H)) (Pierce Chemical Co.)
- PBS (see *Protocol 4*)
- Acetate buffer: 0.1 M sodium acetate, 100 mM sodium chloride pH 4.5

Method

1. Prepare a 3 ml solution containing 4 mg of antibody in cold reaction buffer.

2. Add 300 μl cold sodium meta-periodate solution (25 mg/ml in reaction buffer) to the antibody solution.

3. Incubate the mixture for 1 h in the dark at room temperature.

4. To stop the reaction, add 5 μl glycerol, mix thoroughly, and incubate for 5 min.

5. Divide the antibody solution into microfilterfuge tubes and spin in a microcentrifuge for 15 min at 8000 r.p.m. This is to remove the sodium meta-periodate.

6. Bring the antibody sample up to its original volume with acetate buffer.

7. Repeat steps 5 and 6 two more times.

8. Dissolve the carbohydrate reactive cross-linker (10 mg/ml) in DMSO.

9. To the 3 ml of periodate reacted antibody sample, add 100 μl of the cross-linker and mix thoroughly. Incubate this solution for 2 h in the dark at room temperature while mixing on a shaker.

10. After 2 h, divide the antibody–cross-linker sample into microfilterfuge tubes and spin in a microcentrifuge for 15 min at 8000 r.p.m. This is to remove unreacted cross-linker.

11. Bring the antibody–cross-linker up to its original volume (3 ml) with PBS.

12. Repeat steps 10 and 11 two more times.

13. Determine the concentration of the antibody–cross-linker conjugate by UV–Vis spectrophotometry or other protein determination methods. Bring the sample up to a final concentration of 0.05 mg/ml with PBS.

14. Prepare thiol terminal silane coated solid supports following *Protocol 4*, steps 1 and 2.

15. To the silanized solid supports, add the antibody–cross-linker conjugate and coat the supports thoroughly. Incubate this mixture for 1 h at room temperature.

16. Rinse the antibody coated supports with PBS containing 0.1% Tween 20 and store in PBS at 4 °C.

It should be noted that determination of the antibody–cross-linker conjugate concentration by UV–Vis spectroscopy is not straightforward, as the carbohydrate cross-linkers also absorb at 280 nm. For the maleimide containing cross-linkers, the 280 nm reading can be used to estimate the antibody concentration. For the pyridyl disulfide cross-linker, half of the reading at 280 nm can be used to estimate antibody concentration. Using this protocol, antibody densities of 1.6 ng/mm^2 with an antigen/antibody mole ratio of 0.37 have been obtained on glass substrates.

5. Immobilization procedures using amino-terminal silanes

In procedures employing the amino-terminal silanes, the biomolecule can be attached through primary amines or sulfhydryls. If primary amines are used as the biomolecule's reactive groups, homobifunctional cross-linkers are then employed. Caution needs to be taken to prevent extensive cross-linking, which can affect biomolecule function, when using homobifunctional cross-linkers. To be able to use heterobifunctional cross-linkers, the biomolecule must have free sulfhydryls or be modified such that the thiols are available. Modifying proteins usually has an adverse effect on activity or function. The yield of active modified proteins is reduced.

The most common method for immobilizing biomolecules onto amino silane surfaces employs glutaraldehyde. Glutaraldehyde is a homobifunctional cross-linking agent. Several studies have demonstrated that the commercial aqueous solutions of glutaraldehyde are a mixture of mono and multimer structures (20). These different structures all react with proteins in different ways which may affect the degree and type of immobilization. Due to the nature of glutaraldehyde, Walt has recommended that protocols be written specifically for each protein and application (20). *Protocol 6* was compiled from three articles reporting on the immobilization of antibodies to solid supports with amino-terminal silane and glutaraldehyde (12, 13, 21).

Protocol 6. Protein immobilization with aminopropyltriethoxysilane and glutaraldehyde

Equipment and reagents

- Vacuum oven
- 3-aminopropyltriethoxysilane (APTS) (Aldrich, Sigma)
- Absolute ethanol
- Glutaraldehyde, E.M. grade (Sigma, Poly-Sciences)
- PBS (*Protocol 4*)

- Carbonate buffer: 0.1 M sodium carbonate pH 9.2
- Glycine
- 0.2 M ethanolamine
- Antibody solution: 0.5 mg/ml in carbonate buffer

Protocol 6. *Continued*

Method

1. Immerse the solid support in 5% APTS in distilled water for 30 min at room temperature.

2. Rinse the silanized support thoroughly with distilled water and absolute ethanol sequentially.

3. The support is then cured in a vacuum oven at 80°C overnight.

4. Immerse the silanized support in a 2.5% glutaraldehyde solution in carbonate buffer for 2 h.

5. Rinse the treated support thoroughly with PBS.

6. Immerse the modified support in a 0.5–0.6 mg/ml antibody solution for 7–8 h.

7. Rinse the antibody coated support thoroughly with PBS.

8. To block the unreacted aldehyde groups, the antibody coated support is immersed in either 0.2 M ethanolamine or 0.1 M glycine for 1 h.

9. The final product is rinsed thoroughly with distilled water and stored in PBS.

Another protocol employing an amino-terminal silane uses a carbodiimide to link the biomolecule to the silanized support (22). This procedure is often used when the biomolecule needs to be coupled at an acidic pH to maintain activity. With carbodiimides, no bridge is formed between the silane on the surface and the carboxyl groups on the biomolecule. Some biomolecules may not have the freedom to exhibit functionality without a bridge to extend it away from the surface.

Protocol 7. Protein immobilization with aminopropyltriethoxysilane and carbodiimide

Equipment and reagents

- Water-bath (75°C)
- Oven (115°C)
- Aminopropyltriethoxysilane (APTS)
- Hydrochloric acid (HCl)

- Phosphate buffer: 0.03 M H_3PO_4 pH 4
- 1-cyclohexyl-3-(2-morpholinoethyl)carbodiimide (CMEC)
- PBS (*Protocol 4*)

Method

1. Immerse the cleaned solid support into a 10% (v/v) APTS in distilled water.

2. Adjust the pH of the silane solution to between pH 3–4 with 6 M HCl.

3. Place the silane/support mixture in a water-bath at 75°C for 2 h.

4. Rinse the solid support thoroughly with distilled water.

5. Dry the silanized support in an oven at 115°C for a minimum of 4 h.

6. Place the silanized support in 0.03 M H_3PO_4 buffer and add 2–4 mg/ml CMEC. Mix the solution thoroughly.

7. Add the protein solution (1–2 mg/ml) to the CDI/support solution, mix, and incubate overnight at 4°C.

8. Rinse the protein coated support and store in PBS.

6. Immobilization procedures using epoxy-terminal silanes

The third silane, glycidoxypropyltrimethoxysilane, is occasionally used for protein attachment but not as often as the first two silanes. For attachment protocols employing epoxy-terminal silanes, the epoxide group must be hydrolysed and modified for reaction with the biomolecule directly or with cross-linkers. Once the epoxide is opened, additional modifications can be perform to get the desired reactive group. Pope *et al.* used the following procedure (*Protocol 8*) for the direct immobilization of the biomolecule to the silane surface without the use of cross-linkers (23). Although acidic solutions are the preferred method to open an epoxide ring, use of a basic solution for an extended period obtains the same result.

Protocol 8. Protein immobilization with glycidoxypropyltrimethoxysilane (liquid phase)

Equipment and reagents

- Oven or 1 mm Hg vacuum system
- Glycidoxypropyltrimethoxysilane (GOPS)
- Ethanol
- 0.05 M borate buffer pH 9.6
- Protein
- Phosphate-buffered saline pH 7.4 (PBS)

Method

1. Immerse cleaned substrate in a 10% GOPS solution in 95% ethanol in water for 1 h at room temperature.

2. Remove the silanized substrate from the solution and dry at 50°C for 6 h or under a 1 mm Hg vacuum.

3. Add protein solution (protein in borate buffer pH 9.6) to the silanized substrate and incubate overnight at room temperature.

4. Rinse the protein coated substrate with PBS and store.

Like the earlier silanes, glycidoxypropyltrimethoxysilane has also been used with cross-linkers. One of the major cross-linkers used with this silane is a carbodiimide. Unlike the earlier carbodiimide procedure, the carboxyl group is supplied by the silane and the amino group comes from the biomolecule. This cross-linker does not add any additional molecules between the silane and the biomolecule. If the biomolecule is too close to the surface, activity might be reduced. *Protocol 9* is a modified version of Crowley *et al.* (24) used by Alarie and co-workers to immobilize antibodies onto beads (25).

Protocol 9. Protein immobilization with glycidoxypropyltrimethoxysilane and carbodiimide

Equipment and reagents

- Oven
- Ultrasonic cleaner
- Shaker
- Glycidoxypropyltrimethoxysilane (GOPS)
- Hydrochloric acid (HCl)

- Acetonitrile
- 1,1'-carbonyldiimidazole (CDI)
- PBS (*Protocol 4*)
- Protein: 1–1.7 mg/ml in PBS

Method

1. Immerse the cleaned substrate into a 10% (v/v) GOPS solution in water. Degas the mixture for 10 min with ultrasonic vibration.

2. Heat the mixture to 90°C for 3 h while maintaining the pH at 3 with 1 M HCl to generate a carboxyl group.

3. Rinse the silanized substrate with distilled water thoroughly.

4. Dry the substrate overnight *in vacuo* at 105°C.

5. Immerse the silanized substrate into acetonitrile containing CDI (115 mg/ml). Degas the mixture for 15 min with ultrasonic vibration and vacuum aspiration. Then shake the mixture for 30 min at room temperature.

6. Wash the substrate with acetonitrile thoroughly and air dry. Rinse the substrate with PBS.

7. Immerse the substrate in a protein solution and shake for six days at 4°C.

8. Wash the protein coated substrate three times with PBS and then store.

---

7. Conclusion

There are many procedures for preparing biomolecule-modified surfaces employing silanes with or without cross-linkers. Factors such as biomolecule reactive groups, types of cross-linkers, and intended uses need to be considered when choosing a method for covalent attachment of biomolecules.

Table 1. Protein immobilization comparison by protocol

Protocol	Protein applied	Protein immobilized	Activity	Reference
4	0.05 mg/ml	100–220 ng/cm^2	25–35%	9
5	~ 0.05 mg/ml	160 ng/cm^2	30–35%	18
6	0.6 mg/ml	1.28 μg/cm^2	15%	12
7	1–2 mg/ml	n.d.[a]	n.d.[a]	22
8	1.67 mg/ml	0.5–1 μg/cm^2	n.d.[a]	23
9	1–1.7 mg/ml	12 mg Ab/g beads	1 mol Ag/mol Ab	25

[a] n.d. = not determined.

Irregardless of the immobilization procedure, the surface to which the biomolecules are to be attached must be hydrophilic with reactive hydroxyl groups for effective silane coverage. Cleaning of these surfaces is critical for successful silanization. Several protocols for cleaning solid supports were described. Two of these method required hot basic or acidic solutions which requires extra safety measures to protect the operator. *Protocol 1* employs acids for cleaning but not at elevated temperatures.

Table 1 summarizes examples of each protocol for the ability to couple proteins to a solid support and the activity of the immobilized biomolecule. It is important to note that the concentration of the applied molecule is not of the same magnitude for each protocol. If the biomolecule being immobilized is in limited quantities, *Protocol 4* would permit low concentration of applied protein with still a large percentage of that protein being active. For *Protocol 9*, the area of the beads is unknown, therefore the coverage per cm^2 cannot be calculated for comparison to the other protocols.

Many of the immobilization procedures described are complex, taking several days and many organic chemical steps. Other procedures like the mercapto silane method (*Protocols 4* and *5*) are straightforward, can be performed in a day, and are easy to perform. The amino- and mercapto-terminal silanes are the predominate compounds employed for protein immobilization. In both cases, cross-linkers are employed to covalently attach the biomolecule to the silane.

Two basic types of cross-linkers (hetero- and homobifunctional) are used with these protocols. Heterobifunctional cross-linkers are favoured and are used in two-step reactions which help control the chemistry. Homobifunctional cross-linkers, like glutaraldehyde, have been extensively employed, but the avoidance of unwanted intra- and intermolecular cross-linking is very difficult.

Presented in this article were a few of the methods with which the author is familiar. There are as many other protocols as there are biomolecules. The author hopes that in addition to providing useful procedures for covalently attaching biomolecules to a solid support, factors important in selecting a protocol for your application have also been addressed.

References

1. Ahluwali, A., De Rossi, D., Ristori, C., Schirone, A., and Serra, G. (1991). *Biosen. Bioelect.*, **7**, 207.
2. Ulbrich, R., Golfik, R., and Schellenberger, A. (1991). *Biotech. Bioeng.*, **37**, 280.
3. Klibanov, A.M. (1979). *Anal. Biochem.*, **93**, 1.
4. Ligler, F.S., Shriver-Lake, L.C., Ogert, R.A., Anderson, G.P., and Golden, J.P. (1992). *Proceedings Biosensors'92*.
5. Plueddeman, E.P. (1982). *Silane coupling agents*. Plenum Press, New York.
6. Wong, S.S. (1993). In *Chemistry of protein conjugation and crosslinking*, p 295. CRC Press, Boca Raton.
7. Mattson, G., Conklin, E., Desai, S., Nielander, G., Savage, M.D., and Morgensen, S. (1993). *Mol. Biol. Rep.*, **17**, 167.
8. Brinkley, M. (1992). *Bioconj. Chem.*, **3**, 2.
9. Bhatia, S.K., Shriver-Lake, L.C., Prior, K.J., Georger, J.H., Calvert, J.M., Bredehorst, R., *et al.* (1989). *Anal. Biochem.*, **178**, 408.
10. Jönsson, U., Malmquist, M., Olofsson, G., and Rönnberg, I. (1988). In *Methods in enzymology*, (ed. K. Mosbach), Vol. 137, p. 381. Academic Press, New York.
11. Hong, H-G., Jiang, M., Sligar, S.G., and Bohn, P.W. (1994). *Langmuir*, **10**, 153.
12. Lin, J.N., Heron, J., Andrade, J.D., and Brizgys, M. (1988). *IEEE Trans. Biomed. Eng.*, **35 (6)**, 466.
13. Lin, J.N., Andrade, J.D., and Chang, I-N. (1989). *J. Immunol. Methods*, **125**, 67.
14. Ligler (misspelled Eigler), F.S., Georger, J.H., Bhatia, S.K., Calvert, J., Shriver-Lake, L.C., and Bredehorst, R. (1991). US Patent 5 077 210. December 31, 1991.
15. Bhatia, S.K., Cooney, M.J., Shriver-Lake, L.C., Fare, T.L., and Ligler, F.S. (1991). *Sens. Actuat.*, **3**, 311.
16. Feldman, S.F., Uzgiris, E.E., Penney, C.M., Gui, J.Y., Shu, E.Y., and Stokes, E.B. (1995). *Biosens. Bioelect.*, **10**, 423.
17. Choa, F-S., Shih, M-H., Toppozada, A.R., Block, M., and Eldefrawi, M.E. (1996). *Anal. Lett.*, **29 (1)**, 29.
18. Shriver-Lake, L.C., Donner, B.L., Edelstein, R., Breslin, K., Bhatia, S.K., and Ligler, F.S. (1997). *Biosens. Bioelect.*, **12 (11)**, 1101.
19. Bhatia, S.K., Teixeira, J.L., Anderson, M., Shriver-Lake, L.C., Calvert, J.M., Georger, J.H., *et al.* (1993). *Anal. Biochem.*, **208**, 197.
20. Walt, D.R. and Agayan, V.I. (1994). *Trends Anal. Chem.*, **13 (10)**, 425.
21. Suri, C.R., Raje, M., and Mishra, G.C. (1994). *Biosens. Bioelect.*, **9**, 535.
22. Weetal, H. (1976). In *Methods in enzymology* (ed. K. Mosbach), Vol. 44, p. 134. Academic Press, New York.
23. Pope, N.M., Kulcinski, D.L., Hardwick, A., and Chang, Y-A. (1993). *Bioconj. Chem.*, **4**, 166.
24. Crowley, S.C., Chan, K.C., and Walters, R.R. (1986). *J. Chromatogr.*, **359**, 359.
25. Alarie, J.P., Sepaniak, M.J., and Vo-Dinh, T. (1990). *Anal. Chim. Acta*, **229**, 169.

2

Avidin–biotin immobilization systems

MEIR WILCHEK and EDWARD A. BAYER

1. The avidin–biotin system

Over the past two decades, the avidin–biotin system has evolved to be an extremely useful tool in the fields of biology, biochemistry, and biotechnology. The extraordinary affinity ($K_d \sim 10^{-15}$ M) of avidin (or its bacterial relative, streptavidin) for the vitamin biotin forms the basis for using this system. This approach has found broad application, mainly for localization studies, medical diagnostics, and recombinant genetics. The principles and scope of avidin–biotin technology have been extensively described and catalogued, and the reader is referred to the series of reviews which are available in the literature (1–5) as well as the book which has recently appeared (6).

One field which is currently thriving is the use of the avidin–biotin system for immobilization and/or isolation of biologically active material. In this context, immobilized avidin can be considered a very stable 'pre-activated' carrier which binds biotin-containing molecules with exceptional affinity. For all practical purposes, the bonding between avidin and biotin is irreversible (7).

The 'revival' in this area may be attributed to the improvement and development of a variety of new types of avidin- and streptavidin-based surfaces and methodologies. In particular, the availability of reversible monovalent avidin, or modified avidins, such as nitro-avidin, nitro-streptavidin, and NeutraLite avidin (8), and the imminent breakthrough in the development of mutated forms of avidin and streptavidin (9, 10). Since avidin, streptavidin, their analogues, and their derivatives are all very stable, their immobilization is usually advantageous compared to other proteins. The avidins withstand most chemical and adsorption-based immobilization protocols, and in each case, the affinity and binding capacity are largely retained.

The covalent attachment of a protein is usually the last step in an immobilization procedure. For example, an enzyme is immobilized for a particular type of catalysis, the immobilization of protein A is for the isolation of antibodies, and an immobilized antibody is used for the isolation of a complementary antigen. In contrast, the immobilization of avidin is only the starting point

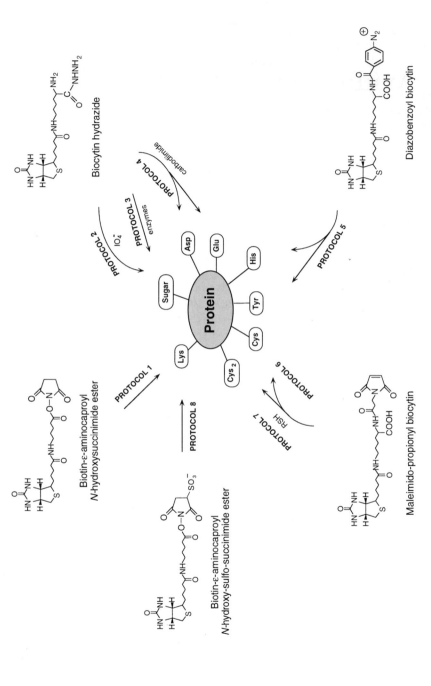

Figure 1. Group-specific biotin-containing reagents for biotinylating proteins. The relevant protocols in this chapter which describe their use are indicated.

of the immobilization procedure. In this context, the avidin surface should be considered as an activated carrier. The only difference is that the avidin-based carrier binds a biotinylated molecule, whereas the chemically activated carrier binds biologically active molecules via different types of covalent bonds. As has been shown by numerous examples, virtually any biologically active compound, including antibodies, receptors, enzymes, inhibitors, hormones, nucleic acids, drugs, and toxins can be easily biotinylated and then bound to the avidin surface (*Figure 1*).

After binding the biotinylated material, the surface can be used for a variety of isolation purposes. Of course, a target molecule can be isolated through interaction with the immobilized biotinylated molecule (the binder), or, alternatively, the immobilized avidin can be used as a simple capture system: to capture the biotinylated binder in complex with its partner (the target) for isolation, e.g. a biotinylated PCR product (11), biotinylated antibody, and CD34$^+$ cells (12). In addition, avidin columns are increasingly being used for the simple retrieval or removal of biotinylated materials from an experimental system. In this context, extraneously applied biotinylated enzymes, antibodies, etc. can be removed once their desired effect has been accomplished (8).

To date, immobilized avidins are currently used for isolation, retrieval, and/or immobilization of receptors, cells, phages, and enzymes (13). Recently, however, one of the major applications involves the immobilization of biotinylated DNA probes and their use for polymerase chain reaction (PCR) (11).

2. Biotinylation of the binder

In order to use this system, one of the interacting partners, the binder, has to be biotinylated to allow for its attachment to the avidin surface for subsequent interaction with the desired target. Throughout the years, different methods for biotinylating various functional groups of proteins, carbohydrates, lipids, and nucleic acids have been described (4, 14). Here, we describe only a few representative examples for biotinylation of proteins (*Protocols 1–7*). Despite the many methods for biotinylating proteins (*Figure 1*), the first method described in the literature, i.e. the use of biotinyl *N*-hydroxysuccinimide ester (BNHS) and similar derivatives, remains the most useful.

Protocol 1.   Biotinylation of proteins on lysine residues

Equipment and reagents

- Dialysis tubing (for retention of M_r 12 000 proteins), boiled in distilled water immediately before use
- Protein: 2 mg/ml dissolved in 0.1 M sodium bicarbonate pH 8.5^a

- Phosphate-buffered saline (PBS) pH 7.4
- Biotinyl *N*-hydroxysuccinimide ester (BNHS) or biotinyl ε-aminocaproyl *N*-hydroxy-succinimide ester (BxNHS) (Sigma, Pierce, Boehringer Mannheim, Molecular Probes, Calbiochem)

Protocol 1. *Continued*

Method

1. Immediately before use, dissolve 2 mg/ml BNHS or 2.7 mg/ml BxNHS in dimethylformamide (heat under a stream of hot tap-water if necessary).

2. For every millilitre of protein solution, add 25 μl of BNHS or BxNHS.

3. Let solution incubate at room temperature for 1 h.

4. Dialyse exhaustively at 4°C (100–1000 volumes, three to five buffer changes) against PBS or other suitable buffer.

5. Store in aliquots at –20°C or under conditions consistent with the stability properties of the protein.

[a] PBS can be substituted for bicarbonate. Do not dissolve in amine-containing buffers, e.g. Tris, ethanolamine.

Protocol 2. Periodate-mediated biotinylation of glycoproteins on the oligosaccharide component

Equipment and reagents

- Dialysis tubing (see *Protocol 1*)
- Glycoprotein: 2 mg/ml dissolved in PBS
- PBS pH 7.4
- 0.1 M sodium periodate: freshly prepared aqueous solution
- Biotin hydrazide (BHZ), biotinyl ε-amino-caproyl hydrazide (BxHZ), or biocytin hydrazide (BCHZ)[a] (Sigma, Pierce, Boehringer Mannheim, Molecular Probes, Calbiochem)

Method

1. Dissolve 10 mg/ml of the biotinylating reagent in PBS (heat under a stream of hot tap-water if necessary).

2. For every millilitre of glycoprotein solution, add 110 μl of the periodate solution.

3. Let solution incubate at room temperature for 30 min.

4. Dialyse for 4 h at room temperature against 2 litres of PBS.

5. Add 250 μl of the biotin reagent per millilitre of protein.

6. Incubate at room temperature for 1 h.

7. Dialyse exhaustively at 4°C (100–1000 volumes, three to five buffer changes) against PBS or other suitable buffer.

8. Store in aliquots at –20°C or under conditions consistent with the stability properties of the protein.

[a] BCHZ is the preferred reagent since it is a highly soluble long chained biotin derivative.

Protocol 3. Enzyme-mediated biotinylation of glycoproteins on the oligosaccharide component

Equipment and reagents

- Dialysis tubing (see *Protocol 1*)
- Glycoprotein: 2 mg/ml dissolved in PBS
- Galactose oxidase from *Dactylium dendroides* (Sigma): 100 U/ml in PBS
- Biocytin hydrazide (BCHZ):[a,b] 20 mg/ml in PBS (heat under a stream of hot tap-water if necessary)

- Sialidase from *Vibrio cholerae* (Behringwerke): 1 U/ml
- PBS pH 7.4, supplemented with 1 mM $CaCl_2$ and 1 mM $MgCl_2$
- 0.15 M NaCl, containing 10 mM EDTA and 0.1% sodium azide

Method

1. To 1 ml of glycoprotein solution, add, in sequential fashion and in the following order: 30 μl each of sialidase, galactose oxidase, and BCHZ solution.

2. Let reaction incubate for 2 h at 37°C.

3. Dialyse exhaustively at 4°C (100–1000 volumes, three to five buffer changes) against NaCl solution.

4. Store in aliquots at −20°C or under conditions consistent with the stability properties of the protein.

[a] BCHZ is the preferred reagent since it is a highly soluble long chained biotin derivative. The other analogues would be difficult to dissolve at such high concentrations.
[b] BCHZ is available from Sigma, Pierce, Boehringer Mannheim, Molecular Probes, and Calbiochem.

Protocol 4. Biotinylation of proteins on aspartate and glutamate residues

Equipment and reagents

- Dialysis tubing (see *Protocol 1*)
- Biocytin hydrazide (BCHZ)[a,b] *N*-(5-aminopentyl)biotinamide:[c] 20 mg/ml in distilled water (heat under a stream of hot tap-water if necessary), brought to pH 5 with 1 M HCl
- Protein

- Water soluble carbodiimide[d]
- PBS pH 7.4
- 0.5 M HCl
- 0.15 M NaCl

Method

1. Dissolve 1 mg protein directly into 1 ml of BCHZ solution.

2. Add 2 mg of solid carbodiimide.

3. Incubate at room temperature.

4. Maintain solution at pH 5 by adding HCl dropwise.

Protocol 4. *Continued*

5. After 6 h, dialyse overnight at 4°C (100–1000 volumes, three to five buffer changes) against PBS.

[a] BCHZ is the preferred reagent since it is a highly soluble long chained biotin derivative. The other analogues would be difficult to dissolve at such high concentrations.
[b] BCHZ is available from Sigma, Pierce, Boehringer Mannheim, Molecular Probes, and Calbiochem.
[c] Biotin cadaverine is available from Molecular Probes.
[d] Such as 1-ethyl-3-(3-dimethyl aminopropyl) carbodiimide hydrochloride (EDC).

Protocol 5. Biotinylation of proteins on tyrosine and histidine residues

Equipment and reagents

- Dialysis tubing (see *Protocol 1*)
- *p*-Aminobenzoyl biocytin (this is a stable precursor of diazobenzoyl biocytin (DBB) and can be purchased commercially from Sigma, Pierce, Boehringer Mannheim, Molecular Probes, and Calbiochem)
- Protein: 1 mg/ml dissolved in PBS
- PBS pH 7.4
- 2 M HCl solution (ice-cold)
- NaNO$_2$ solution: 7.7 mg/ml in ice-cold double distilled water
- 1 M NaOH
- 0.1 M borate buffer pH 8.4

A. *Formation of diazobenzoyl biocytin*

1. Dissolve 2 mg of *p*-aminobenzoyl biocytin in 40.7 µl of ice-cold 2 M HCl.

2. Add 40.7 µl of NaNO$_2$ solution.

3. After 5 min at 4°C, add 35 µl of NaOH to terminate reaction.

4. Add 12 µl of the resultant DBB solution to 1 ml of borate buffer for next step.

B. *Biotinylation step*

1. Add 1 ml of DBB solution to 1 ml of protein solution.

2. Let the solution stand without stirring at room temperature for 2 h.

3. Dialyse overnight at 4°C (100–1000 volumes, three to five buffer changes) against PBS.

4. Store under appropriate conditions (under sterile conditions at 4°C or at –20°C.

Water soluble sulfo analogue of the *N*-hydroxysuccinimide reagents are appropriate for biotinylation of cell surface proteins (*Protocol 8*). The negative charge and water solubility of the sulfo reagents serve to prevent its intracellular penetration, although its chemical reactivity is somewhat reduced compared to BNHS.

Protocol 6.   Biotinylation of proteins on cysteine residues[a]

Equipment and reagents
- Dialysis tubing (see *Protocol 1*)
- Maleimidopropionyl biocytin (MPB), or analogous maleimido derivative of biotin (Sigma, Pierce, Boehringer Mannheim, Molecular Probes, Calbiochem)
- Protein: 2 mg/ml dissolved in PBS
- PBS pH 7

Method

1. Prepare 1 mg/ml solution of MPB in PBS.

2. Add 110 μl of the reagent per millilitre of protein.

3. Incubate at room temperature for 2 h.

4. Dialyse exhaustively at 4°C (100–1000 volumes, three to five buffer changes) against PBS or other suitable buffer.

5. Store in aliquots at –20°C or under conditions consistent with the stability properties of the protein.[b]

[a] Successful biotinylation using this method is dependent on the availability of free, surface-exposed cysteine residues on the protein.
[b] Some sulfhydryl proteins are cryo-unstable after biotinylation and freezing disrupts their structure and activity. In such cases, the protein may be sterilized by passage through a suitable filter and the sterilized protein solution can be stored at 4°C.

Protocol 7.   Biotinylation of proteins on cystine residues

Equipment and reagents
- Dialysis tubing (see *Protocol 1*)
- Maleimidopropionyl biocytin (MPB), or analogous maleimido derivative of biotin (Sigma, Pierce, Boehringer Mannheim, Molecular Probes, Calbiochem)
- Protein: 2 mg/ml dissolved in PBS
- Dithiothreitol: 10 mg/ml freshly dissolved in PBS
- PBS pH 7

Method

1. Add 110 μl of the dithiothreitol solution per millilitre of protein.

2. Incubate at room temperature for 2 h.

3. Dialyse overnight at 4°C (1000 volumes, three to five buffer changes) against PBS.[a]

4. Biotinylate newly exposed sulfhydryl groups according to *Protocol 6*.

[a] The thiol-containing reagent can be removed in alternate fashion using a suitable column (e.g. Sephadex G25).

Protocol 8. *In situ* biotinylation of cell surface proteins

Equipment and reagents

- Centrifuge
- Cultured cells, resuspended to 1–2 × 10⁶ cells/ml in PBS[b]
- PBS pH 7.4

- Biotinyl *N*-hydroxy-sulfo-succinimide ester (B-sulfo-NHS) or biotinyl ε-aminocaproyl *N*-hydroxy-sulfo-succinimide ester (Bx-sulfo-NHS) (Pierce)[a]

Method

1. Immediately before use, dissolve 2.6 mg/ml B-sulfo-NHS or 3.3 mg/ml Bx-sulfo-NHS in PBS (heat under a stream of hot tap-water if necessary).

2. For every millilitre of cell suspension, add 25 µl of the reagent, and incubate for 1 h.

3. Centrifuge and wash cells using an appropriate buffer.

[a] The sulfo-*N*-hydroxysuccinimide reagents are soluble in water and appropriate for biotinylation of cell surfaces in the absence of organic solvents. Their action is not as efficient as the regular *N*-hydroxysuccinimide esters, and their application should be limited to target material which is sensitive to organic solvents.
[b] Other buffers can be used instead of PBS, the major condition being a lack of amino-containing compounds, which would compete with the reaction of cell surface groups with the biotinylating reagent.

Protocols 1–3 are appropriate for biotinylation of most antibody preparations. These protocols are especially suitable for biotinylation of the IgG fraction of polyclonal antibodies, in which case the protein concentration can range from 10 µg/ml to 10 mg/ml, and the amount of biotinylating reagent (BNHS or BxNHS) is adjusted accordingly.

For sensitive monoclonal antibody preparations, if *Protocol 1* inactivates binding or specificity, use *Protocol 2* or *3* as described above.

Variations of the simple BNHS procedure are provided for biotinylation of high and low concentrations of purified antibody (*Protocol 9*). In some cases (i.e. for low titres of antiserum), it is advisable to perform the biotinylation reaction directly on the crude serum without purifying the antibody. For this purpose, a modification of the BNHS procedure is also given in *Protocol 9*.

Protocol 9. Biotinylation of antibodies

Equipment and reagents

- Dialysis tubing (see *Protocol 1*)
- Antibody solution: preferably dialysed against 0.1 M sodium bicarbonate
- PBS pH 7.4

- Biotinyl *N*-hydroxysuccinimide ester (BNHS) or biotinyl ε-aminocaproyl *N*-hydroxysuccinimide ester (BxNHS) (Sigma, Pierce, Boehringer Mannheim, Molecular Probes, Calbiochem)

A. *High concentrations of antibody*

1. For 10 mg/ml of IgG, add freshly prepared BNHS or BxNHS solution (0.68 mg or 0.92 mg, respectively, in 0.1 ml of dimethylformamide)[a] and mix gently but thoroughly.
2. Let solution incubate at room temperature for 1 h.
3. Dialyse exhaustively at 4°C (100–1000 volumes, three to five buffer changes) against PBS or other suitable buffer.
4. Store in aliquots at –20°C.

B. *Low concentrations of antibody*

1. For 10 μg/ml of IgG, add 25 μl of freshly prepared BNHS or BxNHS solution (20 or 27 μg, respectively, in dimethylformamide) and mix.
2. Let solution incubate at room temperature for 1 h and continue as above (part A).

C. *Crude antiserum*

1. Dilute serum (9 vol. 0.1 M sodium bicarbonate to 1 vol. whole serum), and, for every millilitre of solution, add freshly prepared BNHS or BxNHS solution (0.68 mg or 0.92 mg, respectively, in 0.1 ml of dimethylformamide)[a] and mix.
2. Let solution incubate at room temperature for 1 h and continue as above (part A).

[a] If necessary, heat under a stream of hot tap-water, in order to dissolve reagent.

Several methodologies are available for biotinylation of nucleic acids. The simplest and most useful are the nick translation and the photoactivation approach. These are described in *Protocols 10* and *11*, respectively.

Protocol 10. Biotinylation of nucleic acids by nick translation

Equipment and reagents

- Ice-bath
- Vortex
- Sephadex G25 column (Pharmacia)
- Double-stranded DNA preparation: 200 μg/ml
- DNase I stock: 1 mg/ml of DNase I in 0.15 M NaCl and 50% glycerol[a]
- DNase I working solution: dilute 1 μl of stock to 10 ml using nick translation buffer containing 50% glycerol

- 0.2 mM biotinylated dNTP stock
- 0.2 mM unlabelled dNTP stock solutions
- Klenow fragment (*E. coli* DNA polymerase I)
- 0.5 M EDTA
- 10 x nick translation buffer: 0.5 M Tris–HCl pH 7.5, 0.1 M magnesium chloride, 80 mM 2-mercaptoethanol, 0.5 mg/ml BSA (nuclease-free)[a]

Method

1. Combine 5 μl of the following solutions: the DNA preparation, each of the unlabelled and biotinylated dNTPs,[b] and the 10 x nick translation buffer. Add water to 44 μl and chill in an ice-bath.

Protocol 10. *Continued*

2. Add 5 µl of the DNase I working solution and mix by vortexing.

3. Add 5 U of the Klenow fragment, and incubate for 1 h at 16°C.

4. Add 2 µl of the EDTA solution to stop the reaction.

5. Separate the biotinylated DNA from the reactants by passing through a Sephadex G25 column. Separate the biotinylated DNA from un-labelled DNA using a nitro-avidin affinity column (*Protocol 19*).

[a] Store in aliquots at –20°C.
[b] For reduced levels of incorporation, add less biotinylated dNTP.

Protocol 11. Biotinylation of nucleic acids using photobiotin

Equipment and reagents

- Vortex
- Centrifuge
- High intensity mercury vapour lamp, 250 W[a]
- Photobiotin (Sigma, Pierce, Molecular Probes, etc.): 1 mg/ml aqueous solution[b]
- DNA or RNA: 1 mg/ml
- Salmon sperm DNA or RNA: 1 mg/ml

- 0.1 mM EDTA pH 8
- 0.1 M Tris–HCl pH 9
- Isobutanol
- 4 M NaCl
- Ethanol
- Dry ice

Method

1. To 10 µl of the nucleic acid sample, add 20 µl of the photobiotin solution.

2. Incubate in ice-bath and expose solution, 10 cm underneath the mercury vapour lamp, for another 15 min.

3. Bring solution to 100 µl with Tris buffer, and add 100 µl of isobutanol.

4. Mix by vortexing and centrifuge.

5. Discard upper organic phase.

6. Repeat steps 4 and 5.

7. Add 50 µg of salmon sperm DNA or unrelated RNA and 0.75 µl of 4 M sodium chloride.

8. Mix well, add 100 µl of ethanol for DNA (125 µl for RNA).

9. Cool sample for 15 min using dry ice (or –20°C overnight).

10. Centrifuge sample, discard solution, collect precipitate, dry sample, and dissolve in 0.1 M EDTA pH 8.[b]

[a] Optimum wavelength: 350 nm.
[b] Store in aliquots at –20°C.

3. Immobilization of avidin to solid supports

The immobilized avidin (i.e. the avidin surface) can be used to isolate native biotin-containing proteins, proteins engineered to contain biotin, as well as biomolecules to which biotin was artificially incorporated. As in all other affinity-based systems, the avidin column is selective and will adsorb only biotin-containing molecules—the binding is far superior to other systems due to its exceptional affinity constant for biotin. Therefore, if an avidin surface is going to be used, there is often no need for prior separation of non-biotiny-lated material (although there may be some non-specific adsorption). Once the biotinylated molecule is bound to the avidin, the affinity surface is remarkably stable and can survive the most rigorous washing procedures.

For additional stability considerations, the immobilized avidin should retain a complement of free binding sites, and the column should not be over-loaded with biotinylated molecules. Only subsaturating amounts of the biotinylated protein or other molecule should be applied, such that a popula-tion of free binding sites will remain. These would then be available to capture and cross-link biotinylated molecules which may disengage from the avidin matrix due to chemical or enzymatic degradation or due to exchange resulting from the low but finite dissociation constant.

There is almost no difference in the method chosen for coupling the avidin to the surface. Here we provide two different methods (*Protocols 12* and *13*) for immobilization of avidin to Sepharose (*Figure 2*). A method for the adsorp-tion of avidins to plastic surfaces is also described (*Protocol 14*). For safety considerations, the cyanogen bromide method (not presented here, but avail-able in many other books and reviews) has, throughout the years, given way to other, more modern, coupling methods. Thus, the cyano-transfer method (*Protocol 12*) effects the same chemistry as cyanogen bromide, but the cyano-transfer reagent is non-volatile and, consequently, less hazardous (15). Another popular method in recent years is the activated carbonate method

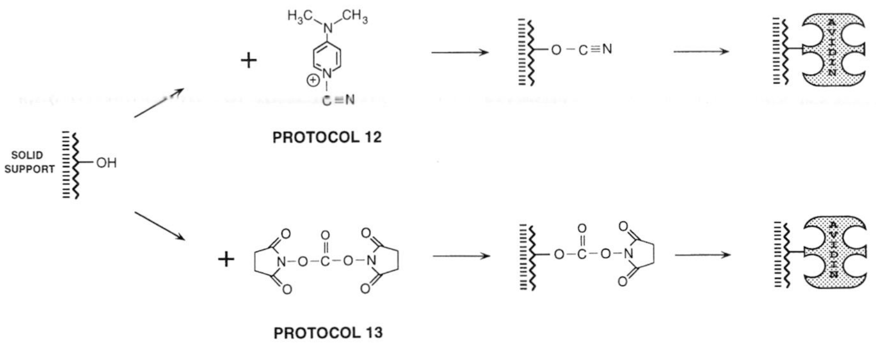

Figure 2. Covalent coupling of avidin to Sepharose using the cyano-transfer method (*Protocol 12*) and the activated carbonate method (*Protocol 13*).

(*Protocol 13*), which results in a more stable covalent bonding of proteins and highly reduced leakage (16). In our hands, the best method has been immobilization using *N*-hydroxysuccinimide derivatives, such as *N,N'*-disuccinimidyl carbonate, *N*-hydroxysuccinimide chloroformate, or *N*-hydroxysuccinimide esters (16). In the case of avidin, even conventional CNBr-mediated coupling has resulted in materials with lower leakage properties than that observed for other proteins. The addition of the biotinylated protein further improves the stability of the matrix, perhaps due to the inherent cross-linking among the avidin molecules by virtue of strong interaction between the tetramer and the target molecules which may bear many biotin groups (17).

Protocol 12.   Preparation of avidin–Sepharose: cyano-transfer method[a]

Equipment and reagents

- Ice-bath
- Sintered glass funnel (coarse)
- Avidin, NeutraLite avidin, streptavidin (SPA), or desired derivative (e.g. nitro-avidin, *Protocol 17*)
- Sepharose CL-4B (Pharmacia Biosystems): 50 g
- Cyano-transfer reagent: *N*-cyanotriethyl-ammonium tetrafluoroborate (Sigma)
- Triethylamine: 0.2 M aqueous solution
- Acetone: 30% and 60% aqueous solutions
- Wash medium: acetone, 0.1 M HCl (1:1)
- PBS pH 7.4

Method

1. Dissolve 50 mg of the desired protein in 50 ml of 0.1 M sodium bicarbonate solution, pH 8.5.

2. Wash the Sepharose by successive filtration in a sintered glass funnel with large volumes (0.5–1 litre) of distilled water, 30% acetone, and 60% acetone.

3. Resuspend Sepharose in 50 ml of 60% acetone, and cool the resin in an ice-bath.

4. Add 500 mg of cyano-transfer reagent.

5. Stir Sepharose (very gently) with a magnetic stirrer, and add dropwise 5 ml of the triethylamine solution.

6. After 2 min, filter the activated gel, wash first with 100 ml of cold wash medium, and then exhaustively with distilled water.

7. Add the avidin solution, and stir overnight at 4°C. Block untreated active groups with a 100 ml solution of 0.1 M glycine or ethanolamine.

8. Wash the gel first with 500 ml of 0.1 M sodium bicarbonate, then with 500 ml of PBS, and resuspend in 50 ml of PBS (containing 0.1% sodium azide). Store at 4°C.

[a] The cyano-transfer method is equivalent to, but much safer than, the classical cyanogen bromide activation method.

Protocol 13. Preparation of avidin–Sepharose: activated carbonate method[a]

Equipment and reagents

- Ice-bath
- Sintered glass funnel (coarse)
- Avidin, NeutraLite avidin, streptavidin (SPA), or derivative (e.g. nitro-avidin, *Protocol 17*): dissolve 50 mg of the desired protein in 50 ml of 0.1 M sodium bicarbonate solution pH 8.5
- Sepharose CL-4B (Pharmacia Biosystems): 50 g

- *N,N'*-disuccinimidyl carbonate (Aldrich)[b]
- Dimethylaminopyridine: 3.25 g in 50 ml acetone
- Dry acetone, methanol, and isopropanol
- Acetone: 25%, 50%, and 75% in aqueous solution
- 5% acetic acid in dioxane
- 0.1 M sodium bicarbonate solution pH 8.5
- PBS pH 7.4

Method

1. Dehydrate 50 g of Sepharose by successive filtration in a sintered glass funnel, using 500 ml volumes of distilled water, graded acetone: water mixtures, and, finally, dry acetone.

2. Resuspend Sepharose in 50 ml of cold dry acetone, containing 4 g of *N,N'*-disuccinimidyl carbonate,[b] and keep the resin in an ice-bath.

3. Stir Sepharose (very gently) with a magnetic stirrer, and add the dimethylaminopyridine solution to the suspension in dropwise fashion.

4. Stir the suspension for 1 h at 4°C.

5. Filter the activated gel, wash first with 500 ml volumes of cold acetone, 5% acetic acid in dioxane, methanol, and isopropanol. Store in isopropanol at 4°C. Before use, wash extensively with distilled water.

6. Add the avidin solution, and stir for 4–16 h at 4°C. Block untreated active groups with a 100 ml solution of 0.1 M glycine or ethanolamine.

7. Wash the gel first with 500 ml of 0.1 M sodium bicarbonate, then with 500 ml of PBS, and resuspend in 50 ml of PBS (containing 0.1% sodium azide). Store at 4°C.

[a] The disuccinimidocarbonate or chloroformate-mediated activation method results in a stabler bond between the protein and matrix than that achieved by the cyanogen bromide or cyanotransfer methods. These stable columns are preferred for long-term usage or for procedures which involve more strenuous conditions for elution.
[b] N-hydroxysuccinimide chloroformate or *p*-nitrophenyl chloroformate (Aldrich, Sigma, or Pierce) can be substituted for *N,N'*-disuccinimidyl carbonate.

Protocol 14. Preparation of avidin- or streptavidin-bound microtitre plates and Petri dishes

Equipment and reagents

- Microtitre plates, Petri dishes
- Avidin, NeutraLite avidin, streptavidin, or desired derivative (e.g. nitro-avidin, *Protocol 17*)

- 0.1 M sodium bicarbonate buffer pH 8.6
- Blocking solution: 2% BSA in PBS, containing 0.05% sodium azide
- PBS

Protocol 14. *Continued*

Method

1. Prepare 10 µg/ml solution of desired protein (avidin, etc.) in bicarbonate buffer.
2. Apply appropriate volume to wells of the plate or dish, and incubate overnight at 4°C.
3. Incubate with blocking solution for 2 h at 23°C.
4. Wash wells with PBS.
5. The plates are ready for use.

Once a biotinylated molecule is stably affixed to a solid support (*Protocol 15*), the molecule can serve as a binder to fish out target molecules from complex biological mixtures (*Protocol 16*). The stable nature of the avidin–biotin bond ensures efficient release of the target material from the matrix by using stringent elution conditions (*Figure 3A*). For this purpose, detergents, chaotropic agents, organic solvents, or other agents can be included in the elution buffer, in order to disengage the target molecule from the binder without affecting the avidin–biotin complex.

Protocol 15. Preparation of affinity surfaces

Equipment and reagents

- Immobilized avidin, NeutraLite avidin, streptavidin (*Protocols 12–14*)[a]
- Any buffer suitable for the stability of the biotinylated binder (e.g. PBS, Tris–HCl, BSA/Tween)
- The biotinylated binder (e.g. biotinylated antibody, enzyme, or other protein, biotinylated nucleic acid, biotinylated cell sample)

Method

1. Add solution or suspension of the biotinylated binder[b] to the avidin-containing surface.
2. Incubate for 30 min at room temperature.[c]
3. Separate unbound material by filtration, centrifugation, settling, or aspiration.
4. Wash material by similar means.
5. The affinity matrix is ready for use.
6. Store under appropriate conditions.

[a] For every milligram of avidin, streptavidin, or derivative, there are approximately 65–75 µmoles of biotin-binding sites (assuming all four binding sites are free of biotin and retain the binding properties after coupling to the column). Use this range as a maximum for determining the potential biotin-binding capacity of a column.
[b] Subsaturating or near-saturating amounts of binder can be applied to the avidin column.
[c] Incubation can be performed at 4°C if desired (e.g. for biotinylated cell samples).

Protocol 16. Affinity purification using avidin– or
streptavidin–Sepharose

Equipment and reagents

- Appropriate dialysis sacs or desalting column
- Immobilized binder (bound to avidin or streptavidin affinity matrix, prepared as in *Protocol 15*)

  Storage buffer: PBS containing 0.1% sodium azide

- Target solution (e.g. if binder is antibody, then antigen-containing solution is the target material)
- Equilibration and wash buffer: PBS
- Elution buffer: 0.1 M glycine–HCl buffer pH 2.2 (for disrupting antigen–antibody interaction)[a]

Method

1. Equilibrate resin with equilibration buffer.

2. Apply target solution to the resin and incubate (either batchwise or arrest the flow of the column) for 30 min.

3. Wash with 10 vol. of wash buffer.

4. Elute the target material with elution buffer.

5. Desalt or dialyse eluted material.[b,c]

6. Regenerate immobilized binder by washing with wash buffer,[a] and store at 4°C in storage buffer.

[a] If desired, buffer can contain detergent, chaotropic agent, or organic solvent.
[b] Check contents and purity of pooled fractions (e.g. using SDS–PAGE) and, where appropriate, combine with other fractions.
[c] Some avidin may be detected in the eluted material, since dissociation of subunits may occur.

---

4. Modified avidins

Recently, derivatives of avidin and streptavidin have been introduced which their biotin-binding property can be reversed quantitatively, using relatively mild conditions. One class of such derivative, includes nitro-avidin and nitro-streptavidin (8), wherein the single binding site tyrosine of either protein is nitrated (*Protocols 17* and *18*). Nitro-avidin or nitro-streptavidin will bind biotin or biotinylated molecules with a very high affinity constant. However, the biotinylated molecule can be exchanged efficiently with free biotin or by using basic conditions (e.g. pH 10 buffers). Thus, a bound biotinylated molecule can be released from a nitro-avidin surface by free biotin, and the avidin can be regenerated effectively using an appropriate basic buffer (*Figure 3B*). Both the biotinylated binder and nitro-avidin matrix can be reconstituted for use in subsequent experiments. Consequently, the nitro-avidin surface can be considered a true 'universal' solid phase, a concept proposed in earlier work (18).

A. PROTOCOL 16: Conventional Avidin Column

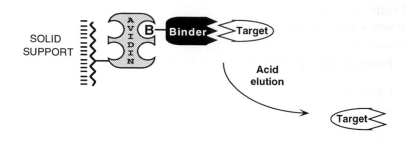

B. PROTOCOL 19: Reversible Nitro-Avidin Column

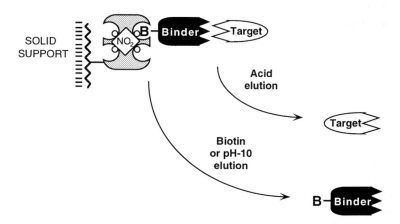

Figure 3. Two types of avidin-containing columns for isolation of target material. (A) The conventional approach (*Protocol 16*) which uses an immobilized form of the native avidin molecule. In this case, the attachment of the biotinylated binder is essentially irreversible, and the original avidin matrix cannot be reconstituted. The stable column can be used for repeated and efficient isolation of target material. (B) The second approach (*Protocol 19*) utilizes new 'reversible' types of avidin, such as nitro-avidin, which retain high levels of affinity for biotin. In this case, target material can be released from the column under one set of conditions, while the biotinylated binder can be released subsequently by using a second set of conditions. In the case of nitro-avidin, the complex with biotin is stable to acidic and neutral conditions up to a pH of about 9. Thus, target material can be released from the binder under acidic conditions, and the binder can later be released from the avidin column upon competition with biotin or by using high pH buffers. Both the biotinylated binder and the regenerated nitro-avidin column can be stored for subsequent use.

The preparation of nitro-avidin (or nitro-streptavidin) is provided in *Protocol 17*. The modified avidin can either be attached directly to a surface (*Protocols 12* and *13*) or immobilized avidin can be nitrated *in situ* (*Protocol 18*). For preparing nitro-avidin plates, it is advisable to adsorb nitro-avidin directly onto the plate (*Protocol 4*), rather than performing the nitration step *in situ*. *Protocol 19* describes the use of a reversible nitro-avidin matrix.

Protocol 17. Preparation of nitro-avidin

Equipment and reagents

- Dialysis tubing (see *Protocol 1*)
- Avidin (Belovo, STC, or SPA),[a] NeutraLite avidin (Belovo), or streptavidin (SPA)
- Tetranitromethane (Fluka)
- 50 mM Tris buffer pH 9.5
- 1 M NaCl

Method

1. Dissolve 5 mg of avidin, NeutraLite avidin, or streptavidin in 1 ml of Tris buffer.

2. Add 2 μl of tetranitromethane for avidin or NeutraLite avidin, or 12 μl of tetranitromethane for streptavidin.

3. Incubate for 30 min at room temperature.

4. Dialyse overnight, once against 4 litres NaCl, and twice against distilled water.

5. Determine the amount of modified tyrosine spectrophotometrically ($\epsilon_{428} = 4200$ Mcm) or by amino acid analysis.

[a] These proteins are also available from retail suppliers, e.g. Sigma or Pierce.

Trial applications of the reversible nitrated matrices have been described (19) and have demonstrated the importance of the potential to dissociate the bond between the immobilized avidin and the biotinylated binder. In another model work involving phage–peptide libraries (20), immobilized nitro-streptavidin was employed for reversible attachment of a biotinylated ligand. Using this system, conventional acidic conditions led to a release of only 60% of the attached phage. The additional 40% could be released using free biotin and/or pH 10 buffer. The interaction between the biotinylated ligand and the latter phage was indeed shown to be of higher affinity than that observed for the phage released by acid treatment.

Protocol 18. *In situ* nitration of avidin or streptavidin matrix

Equipment and reagents

- Sintered glass funnel (coarse)
- Immobilized avidin, NeutraLite avidin, streptavidin, or derivative (*Protocols 12* and *13*)[a]
- Tetranitromethane (Fluka)
- 50 mM Tris buffer pH 9.5
- 1 M NaCl
- PBS

Protocol 18. *Continued*

Method

1. Wash thoroughly 4 ml of avidin-containing resin using pH 9.5 buffer. Resuspend in 4 ml of buffer.

2. Add 6 µl of tetranitromethane for avidin or NeutraLite avidin (36 µl of tetranitromethane for streptavidin).[a]

3. Incubate for 50 min at room temperature.

4. Wash resin exhaustively, first with 1 M NaCl, then double distilled water, and finally PBS.

5. Resuspend in 4 ml of PBS (containing 0.1% sodium azide), and store at 4°C.

---

[a] The volumes of reagent used here are appropriate for resins containing 1 mg/ml of avidin or streptavidin.

Protocol 19. Affinity purification using nitro-avidin or nitro-streptavidin Sepharose

Equipment and reagents

- Appropriate dialysis sacs or desalting column
- Immobilized nitro-avidin or nitro-streptavidin affinity matrix, prepared as in *Protocol 17* or *18*)
- Biotinylated binder solution (e.g. biotinylated antibody, enzyme, or other protein, biotinylated nucleic acid, biotinylated cell sample)
- Target solution (e.g. if binder is antibody, then antigen-containing solution is the target material)

- Equilibration and wash buffer: PBS
- Elution buffer 1: 0.1 M glycine–HCl buffer pH 2.2 (for disrupting antigen–antibody interaction)[a]
- Elution buffer 2: 0.6 mM biotin, dissolved in PBS (for displacing biotinylated binder)
- Elution buffer 3: sodium carbonate buffer pH 10 (for disrupting interaction between nitro-avidin and biotin)
- Storage buffer: PBS containing 0.1% sodium azide

Method

1. Equilibrate matrix with equilibration buffer.

2. Apply binder solution and treat as described in *Protocol 15*.

3. Wash with 10 vol. of wash buffer.

4. Apply target solution and incubate (either batchwise or arrest flow of column) for 30 min.

5. Wash with 10 vol. of wash buffer.

6. Elute the target material with elution buffer 1.

7. Repeat steps 5 and 6 until no protein can be detected in eluted fractions.

8. Pool eluted material.[b,c] Desalt or dialyse, and store under appropriate conditions (e.g. at –20°C).

9. Wash column with wash buffer.[d]

10. Elute the binder from the column using elution buffer 2.

11. Repeat steps 9 and 10 until no protein can be detected in eluted fractions.

12. Pool eluted material.[b,c] Desalt or dialyse and store under appropriate conditions (e.g. at –20°C).

13. Wash column with wash buffer.

14. Elute additional material from the column using elution buffer 3.

15. Repeat steps 13 and 14 until no protein can be detected in eluted fractions.

16. Pool eluted material.[b,c] Desalt or dialyse and store under appropriate conditions (e.g. at –20°C).

17. Regenerate immobilized avidin column by washing with wash buffer,[a] and store at 4°C in storage buffer.

---

[a] If necessary, buffer can contain detergent, chaotropic agent, organic solvent, or other cleansing agent.
[b] Check contents and purity of pooled fractions (e.g. using SDS–PAGE) and, where appropriate, combine with other fractions.
[c] Some avidin may be detected in the eluted material, since dissociation of subunits may occur.
[d] In applications where high pH treatment is not deleterious to the biotinylated binder or where recovery of the latter is not desired, steps 9–12 can be omitted.

The avidin–biotin system was the first protein–ligand system to have been used for immobilization purposes. In recent years, other such systems have also been applied in a similar manner. Most of these have been based on antibody–hapten interactions with the usual benefits and limitations which accompany immunochemical applications. The avidin–biotin system will undoubtedly continue to serve science and industry.

Acknowledgements

This research was supported by grants from the United States–Israel Binational Science Foundation (BSF), Jerusalem, Israel, from the Israel Science Foundation (founded by the Israel Academy of Sciences and Humanities), from MINERVA, Deutsche–Israelische Wissenschaftliche Zusammenarbeit, Germany. Original research in part of this work was funded by the Baxter Healthcare Corporation (Chicago, Illinois).

References

1. Bayer, E.A. and Wilchek, M. (1978). *Trends Biochem. Sci.*, **3**, N237.
2. Bayer, E.A., Skutelsky, E., and Wilchek, M. (1979). In *Methods in enzymology* (ed. D.B. McCormick and L.D. Wright), Vol. 62, p. 308. Academic Press, San Diego.
3. Wilchek, M. and Bayer, E.A. (1984). *Immunol. Today*, **5**, 39.
4. Wilchek, M. and Bayer, E.A. (1988). *Anal. Biochem.*, **171**, 1.
5. Wilchek, M. and Bayer, E.A. (1989). *Trends Biochem. Sci.*, **14**, 408.
6. Wilchek, M. and Bayer, E.A. (ed.) (1990). *Avidin-biotin technology.* Academic Press, San Diego.
7. Wilchek, M. and Bayer, E.A. (1993). In *Immobilised macromolecules: application potentials* (ed. U.B. Sleytr, P. Messner, D. Pum, and M. Sara), Vol. 51. Springer Series in Applied Biology, Springer–Verlag, London.
8. Morag, E., Bayer, E.A., and Wilchek, M. (1996). *Biochem. J.*, **316**, 193.
9. Airenne, K.J., Oker-Blom, C., Marjomäki, V.S., Bayer, E.A., Wilchek, M., and Kulomaa, M.S. (1997). *Protein Expression Purification*, **9**, 100.
10. Sano, T. and Cantor, C.R. (1991). *Biochem. Biophys. Res. Commun.*, **176**, 571.
11. Lundeberg, J., Wahlberg, J., and Uhlen, M. (1991). *BioTechniques*, **10**, 68.
12. Auditore-Hargreaves, K., Heimfeld, S., and Berenson, R.J. (1994). *Bioconjugate Chem.*, **5**, 287.
13. Bayer, E.A. and Wilchek, M. (1994). In *Egg uses and processing technologies* (ed. J.S. Sim and S. Nakai), Vol. 158. CAB International, Wallingford, UK.
14. Bayer, E.A. and Wilchek, M. (1990). In *Methods in enzymology* (ed. M. Wilchek and E.A. Bayer), Vol. 184, p. 138. Academic Press, San Diego.
15. Kohn, J. and Wilchek, M. (1984). *Appl. Biochem. Biotechnol.*, **9**, 285.
16. Wilchek, M. and Miron, T. (1985). *Appl. Biochem. Biotechnol.*, **11**, 191.
17. Bayer, E.A. and Wilchek, M. (1990). *J. Mol. Recog.*, **3**, 102.
18. Wilchek, M. and Bayer, E.A. (1989). In *Protein recognition of immobilized ligands* (ed. T.W. Hutchens), Vol. 83. Alan R. Liss, Inc.
19. Morag, E., Bayer, E.A., and Wilchek, M. (1996). *Anal. Biochem.*, **243**, 257.
20. Balass, M.E.M., Bayer, E.A., Fuchs, S., Wilchek, M., and Katchalski-Katzir, E. (1996). *Anal. Biochem.*, **243**, 264.

3

Antibodies as immobilization reagents

KAMAL L. EGODAGE and GEORGE S. WILSON

1. Introduction

Immobilized antibodies have been used extensively as reagents for the isolation of chemical and biological molecules where there are no naturally occurring ligands. Due to high specificity of the antibodies, the immobilization of specific antigens on different matrices in the required orientation and in high concentration, for use in analytical and biological applications, is possible. The ability to elicit an immune response to a hapten by conjugation to a carrier molecule extends the applicability of antibodies as immobilizing ligands to small molecules. Although polyclonal antibodies have been used as capture matrices, the advent of monoclonal antibody technology has extended and enhanced the use of this approach. Antibodies covalently immobilized to different solid matrices have been used for extraction of biomolecules in plasma (1–3), and for purification of enzymes (4–8), and receptors (9–11). They have also been used as reagents in purification, concentration, and analysis of environmental samples (12, 13). Immobilized antibodies have also been used to prepare immunoreactors for flow injection immunoassays (14–16) and to prepare enzyme reactors (17–19).

Generally, an immobilization reagent should have the ability to immobilize the molecule of interest in high concentration and proper orientation for further use. When antibodies are used as immobilization reagents for biomolecules, antibodies should be immobilized in high concentration on the support in the proper orientation to obtain the highest possible capture capacity of the matrix. The antibody molecule contains four different functional groups that can be used for immobilization: ϵ-amino groups of lysine, carbohydrate residues in the Fc region of the antibody, and carboxyl residues and sulfhydryl groups of the fragmented antibodies. Reaction with the N terminal amino residues should be avoided as they are in the antigen binding region. A variety of solid supports with different activated groups are available for the antibody attachment. Activated agarose beads, glass beads, magnetic beads, and membranes are some of the activated solid supports that have been used for immobilization of antibodies. These solid matrices can be chemically modified to couple through a specific reactive group on the antibody molecules. For covalent attachment of the antibody molecule, the matrix can be activated with cyanogen bromide (CNBr), *N*-hydroxysuccinimide ester (20), periodate (21), 1,1′-carbonyldiimidazole (CDI) (22), tresyl chloride (23), 2-fluoro-1-methylpyridinium toluene-4-sulfonate (FMP) (24), or epichlorohydrin (25). Antibodies can also be immobilized using avidin–biotin technology where biotin labelled antibody can be bound irreversibly to avidin which is covalently attached to a solid support (18). The formation of a covalent bond between the antibody and the solid support enhances the stability of the capture matrix compared to adsorption methods, especially when the immobilized antibody matrix is used in flow-through systems or high pressure systems for long periods of time. This is also critical when eluting the bound antigen using detergents or low pH buffers. Specific antibodies can also be immobilized via another antibody which is specific to the Fc portion of the first antibody molecule or through the use of protein A or protein G (26).

2. Methods of antibody immobilization

Although the antibodies produced by different animal species differ, the basic functional groups that are useful for immobilization and the number of existing individual functional groups do not change significantly except for the sulfhydryl groups formed from the reduction of inter-heavy chain disulfide bonds. Depending upon the species from which the antibody or its subclass is derived, the number of inter-heavy chain disulfide bonds can vary from one to three, sometimes more. Thus the number of sulfhydryl groups available may vary. The antibody immobilization methods that have been described in this chapter have taken these variations into account, and it will be shown how these variations should be optimized for each individual antibody. Although there are different solid matrices, we have taken as an example in this chapter immobilization of antibodies to Reacti-Gel HW 65 (Pierce) which is a 30–60 μm particle size carbonyldiimidazole activated co-polymer with an exclusion limit of 5×10^6 daltons (27). These methods can be used with other gels and solid supports with similar functional groups with minimal protocol modifications.

Protocol 1. Antibody immobilization through lysine ε-amino groups

Equipment and reagents

- Sintered glass funnel
- Centrifuge
- Continuous variable speed stirrer
- Antibody solution
- Reacti-Gel HW 65

- BCA assay regents
- 0.1 M Carbonate buffer pH 9
- 2 M Tris
- 0.1 M Phosphate buffer pH 7.4 with 0.05% NaN₃

Method

1. Dialyse 1 ml of antibody (5 mg/ml) (Section 3, *Protocols 14–16*) overnight in 0.1 M carbonate buffer pH 9 (coupling buffer).

2. Transfer 2 ml of gel to a sintered glass funnel and remove acetone from the gel by applying a very low vacuum.

3. Wash the gel according to the manufacturer's instructions to remove acetone from the gel.

4. Immediately prior to use, equilibrate the gel by washing with coupling buffer.

5. Place the gel in a glass test-tube.

6. Centrifuge at 500 *g* for 5 min and remove residual coupling buffer by aspiration.

7. Add dialysed antibody to the gel (from step 1).

8. Tumble resulting gel slurry end-over-end for 18 h at 4°C.

9. Allow the gel to settle and remove the supernatant (do not discard).

10. Transfer the gel to a sintered glass funnel and wash with 2 M Tris buffer to deactivate the unbound active sites of the solid support.

11. Wash the gel with 0.1 M phosphate buffer three times.

12. Store in the same buffer with 0.05% NaN_3 at 4°C.

13. The amount of immobilized antibody can be calculated by measuring the protein remaining in the supernatant in a BCA assay (Pierce) (28), and subtracting the amount remaining from the total amount used for the coupling (method of difference).

2.1 Antibody immobilization through sulfhydryl moiety of the Fab' fragment

Protocol 2. Preparation of iodoacetamide gel

Equipment and reagents

- Sintered glass funnel
- Continuous variable speed stirrer
- UV-Visible spectrophotometer
- 0.5 M Ethylenediamine pH 9.5
- 2,4,6-trinitrobenzenesulfonic acid (TNBS)
- Saturated sodium tetraborate solution
- 0.2 M Sodium chloride solution
- 0.2 M Adipodihydrazide solution
- 0.1 M Iodoacetic acid solution
- N-Hydroxysulfosuccinimide

- 1-Ethyl-3-(3-dimethylaminopropyl)carbo-diimide hydrochloride
- 0.1 M 2-Mercaptoethylamine (2MEA)
- 50 mM Tris, 1 mM EDTA, 0.01% NaN_3 pH 8.5
- 0.1 M Phosphate buffer pH 7.4
- 0.1 M Carbonate buffer pH 9
- 2 mM 4-4'-dithiodipyridine (4-PDS)
- Krebs–Ringer phosphate buffer pH 7.2
- Reacti-Gel HW 65

Protocol 2. *Continued*

Method

1. Transfer 10 ml of Reacti-Gel HW65 to a sintered glass funnel and wash the gel to remove the acetone according to the procedure described in *Protocol 1*, steps 2 and 3.

2. Immediately prior to use, wash the gel with 0.1 M carbonate buffer pH 9, drain, and transfer immediately to 20 ml of 0.5 M ethylenediamine pH 9.5.

3. Mix resulting slurry end-over-end for 8 h at room temperature.

4. Transfer the slurry to a sintered glass funnel and wash the gel with 0.1 M phosphate buffer pH 7.4, until washings are free from amines. 2,4,6-trinitrobenzenesulfonic acid (TNBS) (Sigma) can be used to confirm the removal of amines and the number of amino groups reacted with the gel (29).

5. To identify unreacted ethylenediamine (30):
    (a) Add 0.03 M TNBS (25 μl) to 1 ml of washings and ensure complete mixing.
    (b) Allow to stand for 30 min at room temperature (25 °C).
    (c) Read absorbance at 420 nm using 0.03 M TNBS in 0.1 M phosphate buffer pH 7.4 as a reference.

6. To determine the number of amines coupled to the gel:
    (a) Suspend samples of washed gel (200 mg) in 1 ml of 1% TNBS.
    (b) Add 1 ml of saturated sodium tetraborate.
    (c) Stir for 30 min at room temperature and wash with 23 ml of 0.2 M NaCl.
    (d) Incubate 100 μl of washing aliquot with 1 ml of 0.2 M adipodihydrazide and 0.9 ml of saturated borate for 15–20 min.
    (e) Read absorbance at 500 nm ($\epsilon = 16\,500$) and calculate unreacted TNBS by subtraction from its initial concentration.

7. Drain all phosphate buffer from the gel by applying a low vacuum.

8. Transfer modified gel to a container which contains 20 ml of 0.1 M iodoacetic acid (Sigma), 0.1 M 1-ethyl-3-(3-dimethylaminopropyl) carbodiimide hydrochloride (Sigma), and 50 mM *N*-hydroxysulfosuccinimide (Pierce).

9. Tumble slurry end-over-end for 24 h at room temperature.

10. Wash the gel with 200 ml of 0.1 M phosphate buffer pH 7.4, as described above, to remove all unbound materials and store the gel in the same buffer at 4 °C.

11. To determine the number of iodoacetamide groups on the gel:
    (a) React 0.5 ml of iodoacetamide gel with 2 ml of 0.1 M 2-mercaptoethylamine (2-MEA) solution in Tris buffer (50 mM Tris–hydroxymethylaminomethane, 1 mM EDTA, 0.01% NaN$_3$ pH 8.5).

(b) Carefully transfer the gel slurry to a disposable column and wash with 15 ml of the coupling buffer.

(c) Dilute 0.75 ml of washings with 0.75 ml of Krebs–Ringer phosphate buffer pH 7.2.

(d) Mix with 1.5 ml of 2 mM 4,4'-dithiodipyridine (4-PDS) in Krebs–Ringer phosphate buffer pH 7.2.

(e) The amount of 2-MEA coupled can be calculated by the method of difference using the absorbance at 324 nm ($\epsilon = 19\,800$) (see *Protocol 1*, step 13).

2.2 Preparation of antibody Fab' fragments

To prepare Fab' fragments, the antibody should be digested first using the proteolytic enzyme pepsin to cleave the Fc fragment yielding $F(ab')_2$. This proteolytic digestion is carried out using an immobilized pepsin column which affords much better control of product formation than the corresponding solution phase reaction (31).

Protocol 3. Immobilization of pepsin

Equipment and reagents

- Sintered glass funnel
- Continuous variable speed stirrer
- Jacketed column (1 × 10 cm)
- Pepsin (from porcine mucosa)
- AminoLink coupling gel

- 0.1 M Acetate buffer pH 4 containing 0.05% NaN_3
- 1 M Sodium cyanoborohydride
- 0.1 M Glycine–HCL buffer pH 3.6
- BCA reagents

Method

1. Add 4 ml of pepsin (5 mg/ml, from porcine mucosa, 2 × crystallized, 3200–4500 U/mg, Sigma) to 4 ml of gel slurry (AminoLink® coupling gel, 4% cross-linked agarose with aldehyde functionality, Pierce) in 0.1 M acetate buffer pH 4 containing 0.05% NaN_3.

2. In a fume-hood, add 0.2 ml of 1 M sodium cyanoborohydride to the slurry.

3. Tumble end-over-end for 2 h at room temperature.

4. Stop mixing and allow the reaction to proceed for another 4 h.

5. Drain the supernatant through a sintered glass filter using a very low vacuum and collect the fraction for protein analysis by the BCA assay (see *Protocol 1*, step 13).

6. Block unreacted coupling sites by reacting with 0.1 M glycine–HCl buffer pH 3.6.

7. Pack the gel in a jacketed column (1 × 10 cm, Kontes) and store in 0.1 M acetate buffer pH 4 containing 0.05% NaN_3 at 4°C.

Protocol 4. Preparation of F(ab')$_2$ using an immobilized pepsin column

Equipment and reagents

- Temperature-controlled water bath (37°C)
- Peristaltic pump
- HPLC system
- Zorbax G-250 column (HPLC size exclusion column)
- Antibody solution
- 0.1 M Acetate buffer pH 4
- 0.1 M Phosphate buffer pH 7.4 containing 0.01% NaN$_3$

Method

1. Dialyse 1 ml of antibody (20 mg/ml) overnight into 0.1 M acetate buffer pH 4.

2. Equilibrate the pepsin column (2 ml of gel) with the same buffer and allow the column to attain 37°C by connecting to a temperature con-trolled water-bath.

3. Introduce the antibody sample and allow the sample to circulate at a rate of 0.75 ml/min using a peristaltic pump where the total volume of the system is about 5 ml.

4. Withdraw 25 μl samples at 2 h intervals and analyse using HPLC-SEC (size exclusion chromatography, Zorbax G-250, 250 × 9.4 mm, DuPont) to monitor the completion of digestion (M_r IgG = 160 000, F(ab')$_2$ = 110 000).

5. When the digestion is completed, dialyse the circulated solution (molecular weight cut-off 13 500) overnight in 0.1 M phosphate buffer pH 7.4 containing 0.01% NaN$_3$ to remove the smaller molecular weight fragments.

Protocol 5. Preparation of Fab' from F(ab')$_2$[a]

Equipment and reagents

- HPLC system
- Zorbax G-250 column (HPLC size exclusion column)
- Peristaltic pump
- Antibody F(ab')$_2$ solution
- 2-Mercaptoethylamine (2 MEA)
- 300 mM iodoacetamide solution
- G25 column (1 × 30 cm)
- 50 mM Tris containing 1 mM EDTA pH 8.5
- 0.1 M Phosphate buffer pH 6
- 1 M Carbonate buffer pH 9.35

Method

1. Dialyse 1 ml of F(ab')$_2$ (5 mg/ml) fragments overnight against 0.1 M phosphate buffer pH 6.

2. Prepare stock solution of 100 mM 2-MEA in 0.1 M phosphate buffer pH 6.

3. Add 2-MEA to the $F(ab')_2$ solution to obtain a final concentration of 50 mM.

4. Withdraw 200 μl samples from the reaction mixture at 30 min time intervals, and block the thiol groups by adding 25 μl of a 300 mM iodoacetamide (Sigma) solution.

   (a) Raise pH to 8.5 using 20 μl of 1 M carbonate buffer pH 9.35.

   (b) Analyse the sample by HPLC-SEC to determine the progression of reaction (M_r Fab' = 55 000).

5. When the reaction is complete, pass the sample through a G25 column (1 cm × 30 cm, flow rate 0.85 ml/min) previously equilibrated with 50 mM Tris–aminomethane, 1 mM ethylenediaminetetraacetic acid (EDTA) pH 8.5, to stop the reaction process by separating the reaction components. These Fab' fragments can be directly used in the coupling reaction (*Protocol 6*). Reformation of the $F(ab')_2$ fragments can be reduced by adding 1 mM EDTA to the separation buffer.

6. The number of sulfhydryl groups can be determined by a 4-PDS assay (see *Protocol 2*, step 11c and 11d).

[a] The conditions for the reaction must be carefully optimized as they can be quite different even for antibodies prepared to the same antigen in the same animal species. The procedure will also depend on the number of inter-heavy chain disulfide bonds.

Protocol 6. Immobilization of Fab' on activated iodoacetamide gel

Equipment and reagents

- Continuous variable speed stirrer
- Antibody Fab' solution
- Iodoacetamide gel (*Protocol 2*)
- BCA assay reagents

- 50 mM Tris buffer containing 1 mM EDTA pH 8.5
- 0.1 M Phosphate buffer pH 7.4 containing 0.01% NaN_3

Method

1. Equilibrate 1 ml of iodoacetamide gel in Tris buffer (50 mM Tris–aminomethane, 1 mM EDTA pH 8.5).

2. Add 4 ml of Fab' fragments (0.5 mg/ml) in the same Tris buffer (optimum activity; Section 3, *Protocols 14–16*).

3. Carry out the coupling reaction by tumbling end over end for 6 h at 4°C.

4. Wash the gel with 0.1 M phosphate buffer pH 7.4 and store the gel in the same buffer containing 0.01% (w/v) NaN_3.

5. The coupling efficiency can be studied using the BCA assay for the supernatant (see *Protocol 1*, step 13).

2.3 Antibody coupling through carbohydrate moieties

Protocol 7. Preparation of hydrazide gel

Equipment and reagents

- Continuous variable speed stirrer
- Sintered glass funnel
- Reacti-Gel HW 65
- 0.1 M Phosphate buffer pH7

- Adipic dihydrazide
- 0.1 M Phosphate buffer pH 7 containing 0.05% NaN_3

Method

1. Equilibrate 10 ml of Reacti-Gel HW 65 as described above (*Protocol 1,* steps 2 and 3).

2. Immediately before the reaction wash the gel with pH 7 phosphate buffer and drain.

3. Transfer the gel to a reaction vessel and add 20 ml of 0.5 M adipic dihydrazide (Aldrich) prepared in 0.1 M phosphate buffer pH 7.

4. Tumble the reaction mixture end-over-end for 6 h at room temperature.

5. Transfer the slurry to a sintered glass funnel and wash with 0.1 M phosphate buffer until the gel is free of adipic dihydrazide.

6. TNBS can be used to confirm the presence of adipic dihydrazide in the wash (*Protocol 2*, step 5).

7. Store the gel in 0.1 M phosphate buffer containing 0.05% NaN_3.

8. The number of hydrazide groups attached can be calculated by the Wilchek method (29) (*Protocol 2*, step 5).

Protocol 8. Oxidation of the carbohydrate moiety of the antibody molecule

Equipment and reagents

- Peristaltic pump
- UV–visible spectrophotometer
- Antibody solution

- 0.1 M Acetate buffer containing 0.15 M sodium chloride and 0.05% NaN_3 pH 5.5
- 0.2 M Sodium metaperiodate
- G25 column (1 × 30 cm)

Method

1. Dialyse 1 ml of IgG (13 mg/ml) overnight at 4°C against acetate buffer (0.1 M sodium acetate–acetic acid, 0.15 M NaCl, 0.01% NaN_3 pH 5.5).

2. Oxidize the carbohydrate moiety by adding 0.2 M sodium meta-periodate to a final concentration of 10 mM.

3. The oxidation reaction should be carried out in the dark, at room temperature, for 30 min.

4. Stop the oxidation reaction by passing through an equilibrated G25 column (30 cm × 1 cm) eluting with acetate buffer at 1 ml/min.

5. The concentration of the oxidized antibody can be calculated from the absorbance at 280 nm (absorption coefficient 1.42 mg/ml/cm).

6. The binding activity of the oxidized antibodies can be studied by an ELISA method after blocking the aldehyde with 0.1 M Tris buffer pH 8.5, and then passing the solution through an affinity column prepared by immobilization of the respective antigen.

Protocol 9. Immobilization of oxidized IgG molecules

Equipment and reagents

- Continuous variable speed stirrer
- Sintered glass funnel
- Antibody solution (antibody carbohydrate moieties are oxidized)
- Hydrazide activated gel *(Protocol 7)*
- 0.1 M Acetate buffer containing 0.15 M sodium chloride and 0.01% NaN_3 pH 5.5
- 0.1 M Phosphate buffer pH 7.4 containing 0.01% NaN_2
- BCA assay reagents

Method

1. Immediately before the coupling reaction, equilibrate the hydrazide activated gel in acetate buffer pH 5.5 (0.1 M sodium acetate–acetic acid, 0.15 M NaCl, 0.01% NaN_3).

2. Using the optimized antibody concentration (see Section 3, *Protocols 14–16*) add the required amount of antibody for the coupling reaction.

3. Let the coupling reaction proceed for 6 h at room temperature with end-over-end mixing.

4. Transfer the slurry to a sintered glass funnel and remove the unreacted antibodies using acetate buffer.

5. Calculate the coupling efficiency by the method of difference using the BCA assay (see *Protocol 1*, step 13).

6. The coupled gel should be stored in 0.1 M phosphate buffer pH 7.4 containing 0.01% NaN_3 at 4°C.

2.4 Specific antibody immobilization using a capture antibody

Specific antibody molecules can be immobilized to a solid support by first immobilizing a capture antibody which is specific to the Fc portion of the specific antibody (e.g. if the specific antibody is a mouse antibody, then goat anti-mouse Fc-specific antibody is employed) (19). The immobilizing antibody will accordingly bind to an epitope (Fc, which is remote from the specific antibody binding sites).

Protocol 10.   Antibody immobilization through the Fc domain

Equipment and reagents

- Continuous variable speed stirrer
- Sintered glass funnel
- UV–visible spectrophotometer

- FC specific antibody solution
- 0.1 M Phosphate buffer pH 7.4 containing 0.01% NaN_3

Method

1. Immobilize the Fc-specific capture antibody to the support using one of the methods described above.
2. Transfer the required amount of the matrix with capture antibody to a test-tube and remove the buffer by aspiration.
3. Add the specific antibody solution of optimum concentration for the highest activity (see Section 3, *Protocols 14–16*) and mix using end-over-end rotation.
4. Mix the slurry continuously for 2 h at room temperature.
5. Transfer the slurry to a sintered glass funnel and remove the un-reacted antibody by washing with three to four gel volumes of 0.1 M phosphate buffer pH 7.4.
6. Calculate the coupling efficiency by the method of difference using the absorbance at 280 nm (absorption coefficient 1.42 mg/ml/cm).
7. Store the gel in 0.1 M phosphate buffer pH 7.4 containing 0.01% NaN_3.

2.5 Antibody immobilization through an avidin–biotin linkage

Protocol 11.   Coupling of avidin to the solid support

Equipment and reagents

- Sintered glass funnel
- Continuous variable speed stirrer
- Reacti-Gel HW 65
- Avidin solution

- 0.1 M Carbonate buffer pH 9.35
- 2 M Tris buffer pH 8.5
- 0.1 M Phosphate buffer containing 0.01% NaN_3 pH 7.4

Method

1. Transfer the required amount of Reacti-Gel® HW65 to a sintered glass funnel, remove acetone, and wash the gel according to *Protocol 1*, steps 2–3.

---

2. Immediately prior to use, wash the gel with 0.1 M carbonate buffer pH 9.35 and drain the buffer by applying a small vacuum.

3. Immediately transfer the gel to a test-tube.

4. Add avidin (3 mg/ml per ml of gel, Calbiochem) directly dissolved in coupling buffer.

5. Continue the coupling reaction using mechanical inversion for 30 h at 4°C.

6. Transfer the gel slurry to a sintered glass funnel and remove the un-reacted avidin by applying a very low vacuum.

7. Block unreacted active sites by washing with 2 M Tris buffer pH 8.5, and then wash with 0.1 M phosphate buffer pH 7.4 containing 0.01% NaN_3.

8. Store the gel in the same phosphate buffer at 4°C.

Protocol 12.   Antibody labelling with NHS-LC-biotin

Equipment and reagents

- UV–visible spectrophotometer
- Stirring plate
- Peristaltic pump
- Antibody solution
- NHS-LC biotin

- 0.1 M Carbonate buffer pH 8.5
- 0.1 M Phosphate buffer pH 7.4
- 0.1 M Phosphate buffer containing 0.01% NaN_3 pH 7.4
- G25 column (1 × 30 cm)

Method

1. Concentrate the required amount of antibody to a volume of 1–2 ml.

2. Dialyse specific antibodies overnight against 0.1 M carbonate buffer pH 8.5.

3. Transfer the antibody to a test-tube and calculate the antibody concentration using absorption at 280 nm.

4. Add NHS-LC-biotin (15-fold molar excess over antibody) dropwise to a constantly stirred antibody solution. The biotin solution should be prepared in high concentration (ten times the concentration required) with nano-pure H_2O, and should be immediately added to the antibody solution with minimal dilution.

5. The reaction is carried out for 3 h at room temperature (constant stirring).

6. Stop the reaction by passing through a G25 column (1 cm × 30 cm) equilibrated with 0.1 M phosphate buffer pH 7.4 (flow rate 0.85 ml/min).

7. Collect the first elution peak from the column, monitored at 280 nm using a flow cell, or by measuring the absorbance at 280 nm of the collected fractions.

8. Dialyse the biotin labelled antibody in 0.1 M phosphate buffer pH 7.4 containing 0.01% NaN_3 at 4°C, and store at the same temperature.

Protocol 13. Coupling of biotin labelled antibodies to avidin activated support

Equipment and reagents

- Sintered glass funnel
- UV–visible spectrophotometer
- Avidin couple gel (*Protocol 11*)

- Biotin label antibody solution
- 0.1 M Phosphate buffer containing 0.01% NaN₃

Method

1. Transfer 0.5 ml of avidin coupled gel to a sintered glass funnel and drain the buffer used to store the gel.

2. Transfer the gel to a test-tube and mix with 5 ml of biotin labelled antibody (2 mg/ml) (see Section 3, *Protocols 14–16*) in 0.1 M phosphate buffer pH 7.4.

3. Continue the reaction by mechanical inversion for 2 h at room temperature.

4. Transfer the slurry to a sintered glass funnel and apply a small vacuum to drain the unreacted biotin labelled antibody.

5. Collect the supernatant and calculate the coupling efficiency by the method of difference, measuring absorbance at 280 nm.

6. Wash the gel with five gel volumes of 0.1 M phosphate buffer containing 0.01% NaN₃, and store at 4°C.

3. Optimization of antibody immobilization to matrices

Although the manufacturers of different matrices have suggested protein concentrations that should be used in the immobilization reaction, most of these have not been optimized for antibodies or for specific applications. Therefore, it is important to optimize all conditions before actually coupling the antibodies for the specific application. Such optimization studies will help reduce the cost and time required for the preparation of the solid support by reducing the amount of antibody needed and increasing the binding activity of the antibody on the immobilized matrix. In this section, we discuss coupling goat anti-horseradish peroxidase antibodies to tosyl activated Dynabeads M-450 (4.5 μm diameter, Dynal Corp.) as an example. Similar methods can be used to study the optimal immobilization of antibodies to other supports.

Protocol 14. Study of optimal coupling concentration of antibody

Equipment and reagents

- Magnet
- Shaker
- Tosyl-activated Dynabeads M-450
-).1 M Borate buffer pH 9.5
- 0.1 M Phosphate buffer pH 7.4
- 10% (w/v) monoethanolamine
- BCA assay reagents

Method

1. Dialyse antibodies overnight against 0.1 M borate buffer pH 9.5.

2. Mix magnetic particles thoroughly to achieve a homogeneous solution.

3. Remove five 200 μl (8×10^7) aliquots of magnetic particles and place in test-tubes or in a 24-well culture plate (Corning).

4. Prepare antibody solutions of 50, 75, 100, 125, and 150 μg/ml in 0.1 M borate buffer pH 9.5.

5. Transfer 200 μl of each solution to continuously shaking magnetic bead solutions.

6. Allow the reaction to proceed for 24 h at room temperature.

7. Apply a magnetic field to the bottom of a 24-well plate or to the test-tubes and remove the unreacted antibodies.

8. Wash the coupled magnetic particles with 0.1 M phosphate buffer and apply the magnetic field to remove washings. Repeat the washing step three more times.

9. Block unreacted active sites by reacting with a 10% (w/v) solution of monoethanolamine (Fisher) in 0.1 M borate buffer pH 9.5 for 2 h.

10. Wash the magnetic particles with 0.1 M phosphate buffer as described in step 8.

11. Resuspend the antibody immobilized particles in 200 μl of 0.1 M phosphate buffer pH 7.4.

12. Remove a 50 μl fraction from each sample and calculate the amount of antibody coupled by directly carrying out the BCA assay on the protein immobilized on the magnetic particles.

13. Plot the amount of antibody immobilized versus the coupling antibody concentration to determine the maximum concentration of antibody solution that should be used for the coupling reaction.

The amount immobilized versus coupling antibody concentration for goat anti-horse-radish peroxidase antibody is shown in *Figure 1*. This is a typical plot for this type of study and indicates that concentrations higher than

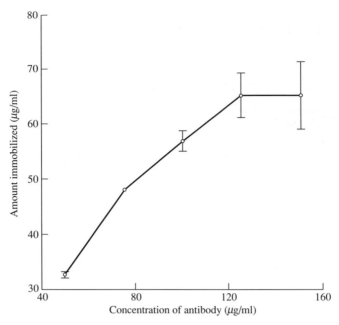

Figure 1. Optimization of the total amount of antibody immobilized on magnetic particles at different immobilization concentrations.

125 µg/ml will not increase the amount of immobilization on the magnetic particles.

Protocol 15. Study of antibody immobilization with coupling time

Equipment and reagents

- Shaker
- Magnet
- Tosyl-activated Dynabeads M-450

- 0.1 M Borate buffer pH 9.5
- 0.1 M Phosphate buffer pH 7.4
- 10% (w/v) monoethanolamine

Method

1. Follow *Protocol 14*, steps 1–3 with four 200 µl aliquots.
2. Prepare a solution of antibody with the lowest concentration, that gives the highest antibody immobilization.
3. Add 200 µl of this solution to each magnetic particle sample.
4. Shake the antibody/magnetic particle mixture at room temperature and take out samples at 30, 90, 180, and 300 min.
5. For each sample that is taken out from the reaction mixture, follow the procedures described in *Protocol 14*, steps 7–12.

3.1 Study of the activity of the immobilized antigen over time

---

Protocol 16.   Immobilization of an enzyme using an antibody

Equipment and reagents

- Plate reader
- Magnetic beads from *Protocol 14* or *15*
- 0.01 M Phosphate buffer with 0.15 M NaCL and 0.2% BSA

- Antigen of interest
- Enzyme substrate if antigen is an enzyme

Method

1. Transfer two 20 μl samples from each sample in Section 3 to test-tubes.
2. Block non-specific adsorption sites with 500 μl blocking buffer (0.01 M phosphate buffer, 0.15 M NaCl, 0.2% BSA) for 1 h at room temperature.
3. Wash with 0.01 M phosphate buffer plus 0.15 M NaCl using the procedure described in *Protocol 14*, step 8.
4. Add enzyme in a suitable stable buffer pH 7.4 (200 μl) at ten times the molar ratio of the highest amount of antibody immobilized on the particles.
5. Allow the binding reaction to proceed for 1 h and wash the unbound enzyme as in step 3.
6. Transfer the magnetic particles to a 24-well culture plate.
7. Add substrate solution (200–500 μl, depending on the activity of the enzyme) and monitor the colorimetric reaction at 5 min time intervals using a plate reader by transferring 50 μl substrate solution to a microtitre plate after applying a magnetic field. Return the transferred solution after the reading.
8. Plot the activity versus immobilization time.

If the immobilized antibody is specific for a protein, then a radiolabelled antigen can be used to study the activity of the solid support. Activity can also be studied using a larger volume of magnetic particles if the absorption co-efficient of the antigen is known and the antibody coated matrix has a high binding capacity. The binding of the antigen can then be measured by disrupting the binding using a low pH buffer and measuring the absorbance at a specific wavelength of the released antigen to estimate the capture capacity of the magnetic particles. If the protein is an enzyme with a low turnover number then a large volume (50 μl) of magnetic beads should be used for the activity study.

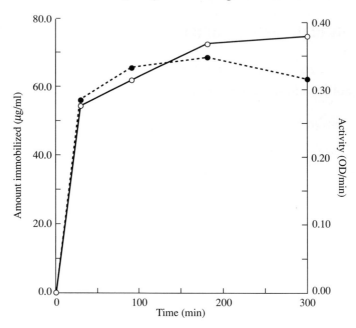

Figure 2. Comparison of the activity with the immobilization time of antibodies. (+) Activity; (●) amount immobilized.

Figure 2 shows plots of the amount of antigen immobilized versus time and activity versus time. In this specific experiment goat anti-horse-radish peroxidase antibodies were immobilized and 20 µl of magnetic particles were used for the activity study (32). This is a typical plot for an activity versus time study. Although the antibody concentration increases with time, the activity is actually reduced after a maximum is attained. This may be due to multiple attachment of the antibody to the surface, or crowding of the immobilized antibody, thus reducing its capture capacity (27).

4. Other factors that are important for the high capture capacity of the immobilized antibodies

4.1 Study of the relationship between capture capacity of immobilized antibodies and the size of antigen

Wimalasena and Wilson (27) have studied the effect of the specific activity of the F(ab′) immobilized on a solid support with the size of the antigen. The results are shown in *Table 1* and *2*. Results from *Table 1* show that when F(ab′) density on the gel is reduced, the biological activity increases to a certain level. This may be due to the low steric hindrance in the F(ab′) fragment when there is a low antibody density. *Table 2* shows that when the size of the

Table 1. Effect of F(ab') density on the specific activity of immobilized goat anti-human F(ab') (27)

Density of F(ab') (mg/ml of gel)	Amount of antigen bound (mg)	Specific activity[a] (%)
4.54	2.20	25
3.43	1.81	28
1.42	1.08	40
1.14	0.92	42

[a] Number of moles of antigen bound/number of moles of antibody bound (average of three determinations); M_r of goat anti-human F(ab') 46 000 and human IgG 160 000.
Goat anti-human F(ab') was coupled to iodoacetamide activated HW 65 at pH 8.5 in 50 mM Tris–HCl buffer. Specific activity was determined under equilibrium conditions with human IgG as the antigen.

Table 2. Specific activity of immobilized goat anti-human F(ab') immunosorbants with a smaller antigen: effects of F(ab') density (27)

Density of F(ab') (mg/ml of gel)	Amount of antigen bound (mg)	Specific activity[a] (%)
4.54	1.26	39
3.43	1.10	45
1.42	0.87	86
1.14	0.82	100

[a] Number of moles of antigen bound/number of moles of antibody bound (average of three determinations); M_r of goat anti-human F(ab') 46 000 and human F(ab') 60 000.
Goat anti-human F(ab') coupled to iodoacetamide activated HW 65 at pH 8.5 in 50 mM Tris–HCl buffer. Specific activity was determined under equilibrium conditions using human F(ab') as the antigen.

antigen is additionally reduced, the specific activity of the immobilized anti-bodies can increase up to 100%. This shows that steric hindrance by the antigen is also a factor in determination of the activity of the immobilized antibodies and thus suggests limits on the effective immobilized antigen density.

4.2 Relationship between selection of antibodies for the immobilization of enzymes for bioanalytical application

Gunaratna and Wilson (19) have discussed the application of antibodies as immobilizing reagents in multienzyme flow reactors in analysis of acetyl-choline. In these studies due to the lower dissociation constant of anti-choline oxidase antibodies compared to the anti-acetylcholinesterase antibodies, reduced lifetime of the multienzyme reactors has been observed due to the leaching of the choline oxidase enzyme from the reactor. Therefore, when antibodies are used as immobilizing reagents in bioanalytical applica-tions where the immobilized antigen is associated with the outcome of the response, high affinity antibodies should be used to reduce the variations of the response due to the leaching of the antigen immobilized during the application.

In these studies, immobilization of enzymes through the avidin–biotin system has generated the best detection limits. This may be due to the lower activity of the choline oxidase enzyme after binding to anti-choline oxidase antibodies. Therefore selecting the antibodies in these types of applications that do not reduce the activity of the enzyme upon antibody binding is important.

5. Conclusion

When antibodies are used as immobilization reagents it is important to optimize all the parameters of the antibody immobilization on the solid support. Although antibody immobilization steps are optimized, the activity of the immobilization reagent will depend on multisite attachment, multiple orientations, and steric hindrance. Multisite attachment and multiple orientation could be studied by the amount immobilized versus time and activity of solid support versus time experiments. Therefore, it is important to understand the balance between these factors when generating the supports with antibodies as immobilization reagents, to achieve the optimal performance.

References

1. Davis, G. C., Hein, M. B., and Chapman, D. A. (1986). *J. Chromatogr.*, **366**, 171.
2. Wojchowski, D. M., Sue, J. M., and Sytkowski, A. (1987). *Biochim. Biophys. Acta*, **913**, 170.
3. van Ginkel, L. A., Stephany, R. W., Rossum, H. J., Steinbuch, H. M., Zomer, G., van de Heeft, E., *et al.* (1989). *J. Chromatogr.*, **489**, 111.
4. Abouakil, N., Rogalska, E., Bonicel, J., and Lombardo, D. (1988). *Biochim. Biophys. Acta*, **961**, 299.
5. Yang, C., Ryu, Y. W., and Ryu, D. D. Y. (1988). *Hybridoma*, **7**, 377.
6. Santos, E., Tahara, S. M., and Kaback, H. R. (1985). *Biochemistry*, **24**, 3006.
7. Brent, T. P., Vonwrouski, M., Pagrane, C. N., and Bigner, D. D. (1990). *Cancer Res.*, **50**, 58.
8. van Faassen, H., van den Berg, I. E., and Berger, R. B. (1990). *J. Biochem. Biophys. Methods*, **20**, 317.
9. Moncharmont, B., Su, J. L., and Parikh, I. (1982). *Biochemistry*, **21**, 6916.
10. Phillips, T. M. and Frantz, S. C. (1988). *J. Chromatogr.*, **444**, 13.
11. Stauber, G. B., Aiyer, R. A., and Aggarwal, B. B. (1988). *J. Biol. Chem.*, **263**, 19098.
12. Thomas, D. H., Beck-Westermeyer, M., and Hage, D. S. (1994). *Anal. Chem.*, **66**, 3823.
13. Wong, R. B., Pont, J. L., Johnson, D. H., Zulalian, J., Chin, T., and Karu, A. E. (1995). *ACS Symp. Ser.*, **586**, 235.
14. De Alwis, W. U. and Wilson, G. S. (1985). *Anal. Chem.*, **57**, 2754.
15. De Alwis, W. U. and Wilson, G. S. (1987). *Anal. Chem.*, **59**, 2786.

16. Lee, J. H. and Meyerhoff, M. E. (1990). *Anal. Chim. Acta*, **239**, 47.
17. De Alwis, W. U., Hill, B. S., Meiklejohn, B. I., and Wilson, G. S. (1987). *Anal. Chem.*, **59**, 2688.
18. De Alwis, W. U. and Wilson, G. S. (1989). *Talanta*, **36**, 249.
19. Gunaratna, P. C. and Wilson, G. S. (1990). *Anal. Chem.*, **62**, 402.
20. Cuatrecasas, P. (1970). *J. Biol. Chem.*, **245**, 3059.
21. Ferrua, B., Maiolini, R., and Masseyeff, R. (1979). *J. Immunol. Methods*, **25**, 49.
22. Bethell, G. S., Ayers, J. S., Hancock, W. S., and Hearn, M. T. W. (1979). *J. Biol. Chem.*, **254**, 2572.
23. Nilsson, K. and Mosbach, K. (1981). *Biochem. Biophys. Res. Commun.*, **102**, 449.
24. Ngo, T. T. (1986). *BioTechnology*, **4**, 134.
25. Matsumoto, I., Mizuno, Y., and Seno, N. (1979). *J. Biochem.*, **85**, 1091.
26. Solomon, B. and Koppel, R. (1987). *Ann. N. Y. Acad. Sci.*, **501**, 463.
27. Wimalasena, R. L. and Wilson, G. S. (1991). *J. Chromatogr.*, **572**, 85.
28. Wiechelman, K., Braun, R., and Fitzpatrick, J. (1988). *Anal. Biochem.*, **175**, 231.
29. Wilchek, M., Miron, T., and Kohn, J. (1988). In *Methods in enzymology* (ed. Jacoby, W. B.), Vol. 104, p. 16. Academic Press, New York.
30. Dean, P. D. G., Johnson, W. S., and Middle, F. A. (1985). In *Affinity chromatography: a practical approach*, p. 76. IRL Press, Oxford.
31. DeSilva, B. S. and Wilson, G. S. (1995). *J. Immunol. Methods*, **188**, 9.
32. Egodage, K. L. (1996). Ph.D. dissertation, University of Kansas.

4

Bioanalytical applications of self-assembled monolayers

BO LIEDBERG and JONATHAN M. COOPER

1. Introduction

The 'self-assembly' of thin molecular films of biological or chemical moieties onto surfaces is a process which is of enormous interest in diverse areas of technology, due to the apparent simplicity by which a solid support can be modified or functionalized. Consequently, there has been a tremendous growth in the interest shown in the use of these systems during recent years, with a recognition of their importance in the development of novel surface coatings. It is now clear that, when the deposition procedure is carefully controlled, it is possible to 'spontaneously' deposit a single layer of particular species (or, indeed, a mixed monolayer of different species) on a surface, giving rise to the use of the term 'self-assembled monolayer' or SAM. Although it is the purpose of this chapter to describe the application of SAMs to bioanalytical methodologies, it should also be acknowledged that the uses of these techniques are numerous and now impinge on many areas of modern materials science. For example, ultrathin SAMs, with predefined compositions have now been used in order to obtain a deeper understanding of interfacial phenomena like adhesion, lubrication, wetting, nucleation, and crystal growth (1, 2).

Within the field of biotechnology, there has been considerable interest in surface modifications (3), with an emphasis being placed on the development of protocols for the attachment of proteins, lipids, and sugars at solid phases (particularly in the production of biological sensors). In achieving this, there has been much effort devoted to controlling the local biological immobilization environment, using suitable surface modifiers in order to change either the physical or the chemical nature of the underlying material. In these circumstances, the SAM may either have the function of performing some aspect of biomolecular recognition in its own right (see *Protocol 4*) or, alternatively, it may act as a 'primer', onto which the biomolecule will be covalently 'grafted' (so producing a bio-functional surface, appropriate for biosensing, *Protocol 5*). In either case, it is important to stress that the amounts of material deposited at the surface are small, and subsequently, devices produced in this manner require considerable care in their preparation, characterization, and application.

In this chapter we describe a series of procedures involving the use of SAMs, which exploit the derivatization either of gold with organosulfur compounds, or of silicon (and or glasses) with silanes (additional protocols for modifying surfaces with silanes are described in Chapter 1). We also show how these methodologies are appropriate for a wide variety of bioanalytical applications, including the development of both optical and electrochemical biosensors. Although, in many cases, these SAM functionalizations could be best performed by deposition from a gaseous phase, for reasons of simplicity, we have limited the experimental description to protocols involving surface modifications from a liquid phase.

2. A historical perspective

When approaching this topic, either as a novice or as a generalist, with some experience of surface functionalization, it is important to realize the full extent of the literature, which is now very large, covering many aspects of the physics, chemistry, and biology of these SAM-modified materials. It is unrealistic to provide a detailed review of this literature in this chapter, although it is, however, relevant to highlight some of the important work, which now underpins this subject.

In general, it is now understood that during 'self-assembly', the species being deposited can either interact predominately with the solid support, and/or with either the solvent or other, neighbouring molecules in solution. For example, there are circumstances when molecular interactions within or between the solvent (cross-linking) will play a more central role in the assembly process than does the direct chemical binding (pinning) of the modifier to the substrate. The most thoroughly studied and well-characterized SAMs which exhibit such 'cross-linking' are those prepared from the silanes (e.g. the alkyltrichlorosilanes including octadecyltrichlorosilane (OTS), which have been deposited onto silica and oxide materials) (4). It is already well-known from many studies that interfacial water controls the extent of cross-linking between silanes and ultimately determines both the degree of their organization and their robustness (4, 5). This point will be explored in more detail later in the chapter (Section 4.3).

The category where pinning is most important is best illustrated by the extensively studied interaction involving organosulfur compounds and gold, using, e.g. alkylthiols, sulfides, and disulfides (1, 2). The structural understanding of the organization of thiol compounds on gold was developed during the early 1980s at Bell Laboratories by Ralph Nuzzo and David Allara (6) although, since then, a variety of other molecules like the alkyl- and arylphosphines have also been shown to be useful modification agents for the noble metals (including the platinum group metals) (7).

The study of the Au–S interaction has shown that, for the *n*-alkanethiols, $(HS(CH_2)_nCH_3)$, $n > 10$, which are the most frequently used compounds in producing model SAM surfaces, the sulfur head groups generally bind as a thiolate, at threefold hollow sites at the Au (111) crystal lattice, forming a commensurate ($\sqrt{3} \times \sqrt{3}$ R = 30°) overlayer structure, with a nearest neighbour distance of about 5 Å (8), as shown in *Figure 1* (upper panel). The strong interaction between the sulfur and the gold substrate contributes substantially to the overall adsorption energy, which is typically in the range 35–45 kcal/mole (145–188 kJ/mole) (8). A slight mismatch between the pinning distance and the van der Waals diameter of the alkyl chain forces the molecules to assemble in a slightly tilted, all *trans* configuration, in order to optimize the lateral interactions, between molecules within the monolayer *Figure 1* (lower panel).

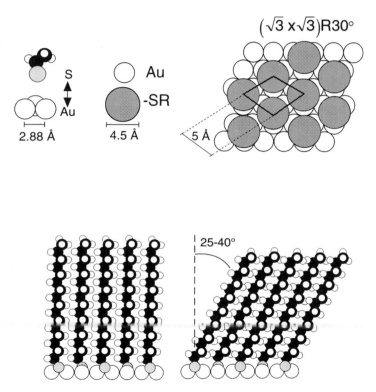

Figure 1. Upper panel shows how the sulfur head groups bind as a thiolate, at threefold hollow sites at the Au (111) crystal lattice, forming a commensurate ($\sqrt{3} \times \sqrt{3}$ R = 30°) overlayer structure, with a nearest neighbour distance of about 5 Å. Lower panel demonstrates how a slight mismatch between the pinning distance and the van der Waals diameter of the alkyl chain forces the molecules to assemble in a slightly tilted, all *trans* configuration, in order to optimize the lateral interactions between molecules within the monolayer.

The degree of crystallinity (all *trans* in nature) is critically dependent on the chain length. As a rule of thumb, alkanethiols with chains shorter than 10–12 methylene units exhibit a rapidly increasing fraction of *gauche* conformers, located at the outermost portion of the alkyl chain, whereas those with longer chains are found to relax in an almost perfect, all *trans* conformation (*Figure 1*). As a consequence the most ordered SAMs are produced from monomer units with longer chain lengths. Where molecular 'organization' is not considered to be of the highest priority, but rather functionality is of greater relevance, then shorter chain length SAMs can be formed.

Whichever type of monolayer is chosen for surface functionalization, whether a silane or a thiol, it is often important to be able to characterize the nature of the surface, paying attention to thickness, coverage, as well as the chemical nature of the species, on the surface. *Table 1* gives a list of techniques which we think are the most important for SAM characterization. As the amounts of material are small, these techniques are necessarily very surface-sensitive. The general analytical capabilities are also indicated and a short description of each technique is given below.

1. Ellipsometry. Ellipsometry is most often used for determining the thickness of a SAM film. The technique measures the change in polarization of light upon reflection (9). The ellipsometric readings delta and phi can be correlated to the thickness and the optical constants of the SAM, e.g. the complex refractive index $N = n + ik$, where n is the refractive index, and k the extinction coefficient. For alkanethiolates, the absorption is normally very weak meaning that k is set to zero and $N = n$. Thus, for a thiol SAM where, $n = 1.5$ (a typical value for an alkyl assembly) and by assuming a three-layer parallel slab model for the gold/SAM/air (or liquid) interface, it is possible to determine the thickness of the film.

2. Contact angle goniometry. The sessile drop method is perhaps the simplest, and at the same time, the most elegant method for characterizing SAMs (10, 11). It is a technique where the contact angle (α) of a probing liquid is measured and correlated to the interfacial tensions of the constituting phases γ_{SL}, γ_{SV}, γ_{LV} through the Youngs Equation (*Equation 1*):

$$\cos \alpha = \frac{\gamma_{SV} - \gamma_{SL}}{\gamma_{LV}} \qquad [1]$$

where γ_{SL} = interfacial tension for the solid liquid interface, γ_{SV} = interfacial tension for the solid vapour interface, and γ_{LV} = interfacial tension for the liquid vapour interface.

The technique can provide information about the chemical nature of the surface, as well about its morphological properties through the hysteresis, $\Delta\alpha = \alpha_a - \alpha_r$, where α_a and α_r are the advancing and receding contact angles, respectively.

Table 1. Analytical capabilities of commonly used techniques for SAM characterization[a]

Experimental technique	Analytical capability							
	Thickness of SAM	Interfacial tension	Coverage	Chemical composition	Orientation of molecule or group	Alkyl chain density	Defects and their distribution	Roughness, chemical homogeneity
Ellipsometry(single wavelength)	++	--	0	--	--	0	--	0
Contact angle goniometry	--	++	-	0	+	0	-	+
Cyclic voltammetry	-	--	++	--	--	++	++	--
Infrared spectroscopy	+	--	+	+	++	++	-	--
X-ray photoelectron spectroscopy	0	--	++	++	+	0	--	--

[a] The table provides a list of techniques which commonly used in the characterization of SAMs. In general XPS and FT-IR provide more quantitative information on the chemical composition of the monolayers and have therefore have been used extensively.
[b] Analytical capability: ++ excellent, + good, 0 fair, - poor, --none.

3. Electrochemistry. The magnitude of the electron transfer between a redox probe in solution and the gold substrate is a good measure of the integrity of the SAM and can be measured using cyclic voltammetry (12). Likewise, impedence analysis can also be used to reveal information about the pin-hole density, and well as about the presence and distribution of sparsely populated regions in the SAM, so-called 'shallow' defects.

4. Infrared spectroscopy provides information about chemical bonds present in a sample and the conformation of the constituent molecules. Due to the dipolar nature of the excitations in the infrared, the technique can also provide information about orientation of molecules and molecular entities using polarized light (13). This feature is of particular importance for SAMs since it can be employed to determine chain and functional group orientation on the surface, a knowledge which can be of considerable importance in considering biological recognition processes.

5. X-ray photoelectron spectroscopy (XPS). This technique provides information about the core electronic levels in atoms and molecules (14). It is the most quantitative method of those described above and has therefore been frequently used to find correlations between the solution and surface (SAM) composition. It can also provide information of the depth of the SAM, by studying the attenuation of the photo-emitted electrons, at different take off angles relative to the substrate surface.

3. Preparation of alkanethiolate SAMs

Protocol 1.   Preparation of the gold substrate, prior to SAM deposition

Reagents

- AL_2O_3
- Metals (Au, Ti, Cr, Pd, etc.)
- SiO_2 substrates
- MPTS
- H_2O_2
- H_2SO_4
- NH_3

Method

1. Wear polythene gloves when handling samples since latex rubber gloves tend to leave behind a small residue of latex particles, or talc, which requires further, extensive, cleaning to remove.

2. The roughness and cleanliness of the supporting substrate material are of vital importance. If polycrystalline gold surfaces are to be used, these can be cleaned by polishing with an alumina (Al_2O_3; Sigma) slurry of decreasing grain size (3 μm, 0.3 μm, and 0.05 μm) in order to minimalize surface roughness. The process should be followed by

---

ultrasonic cleaning in either distilled or reverse osmosis water. The experimentalist may also choose to combine polishing (for the removal of the majority of the dirt) with one or more (wet or dry) etching methods, in order to ensure a higher standard of cleanliness (see below).

3. As an alternative to using polycrystalline gold, substrates can be made relatively cheaply by gold evaporation. The gold should be deposited onto high quality (optically flat) glass, or silicon wafers, in order to avoid substrate-induced roughness and lattice imperfections of the gold film. The gold films should be evaporated or sputtered to a thickness of between 100–200 nm, preferably in ultrahigh vacuum (UHV), but always at a base pressure below 10^{-6} mbar. An oil-free pumping system should be used in order to ensure low contamination levels of the gold film.

4. The adhesion of gold to silicon (SiO_2) is poor. One approach, used in order to promote the adhesion, involves the pre-deposition of a thin (1–10 nm) Cr or Ti underlayer on the substrate. The Ti layer is recommended for high temperature applications because of its lower diffusion constant in Au. In addition, as the Au layer is porous, the solution will tend to interact with the underlayer. Hence, in either case, whether Ti or Cr are used, an intermediate Pd layer can be deposited between the adhesive underlayer and the Au, to act as a diffusion blocking layer. Where sequential deposition of metals is available, a Ti/Pd/Au multilayer of 10/10/100 nm is therefore recommended.

5. Where multilayer deposition of metals is not available, it is recommended that the glass or silicon is first functionalized with a mercaptosilane (MPTS) available from Sigma Chemical Company (see *Protocol 7* for a detailed deposition). The silane of the MPTS interacts with the silicon or the glass, leaving a free thiol to bind Au. This method has the advantage of giving very low 'corrosion' currents, if the substrate is to be used in bio-electroanalysis.

6. Gold films can be prepared in large batches and stored over long periods of time. The organic contamination which inevitably adsorbs on the surface during prolonged storage can be removed by a wet chemical cleaning process in Pirahna solution, comprising a mixture of 1:3 (30% H_2O_2, concentrated H_2SO_4) at 80°C for between 1–5 min.[a]

7. As an alternative method for cleaning gold, it is possible to electrochemically clean the gold surface by electro-oxidation of the metal in 100 mM concentration solutions of either perchloric acid or sulfuric acid at large overpotentials (c. oxidation at > 0.8 V versus Ag|AgCl).

8. A less aggressive, but equally efficient method for cleaning the surface, is to perform the cleaning process in a mixture containing

Protocol 1. *Continued*

30% H_2O_2, 25% NH_3, H_2O (1:1:5) at 80 °C for between 5–10 min. Care should be taken in making up this solution.

9. Following all of these wet 'cleaning' regimes, whether by etching or polishing, copious rinsing in ultrapure distilled water is essential.

10. The best method of cleaning the surface involves the use of one of a variety of other 'dry' techniques, e.g. by using exposure of the surface to ozone, or by etching in an O_2 or an Ar plasma, as illustrated by the XPS spectra for a variety of surface treatments (*Figure 2*). These methods involve the use of expensive laboratory equipment, which will most commonly be found in well equipped physical chemistry, physics, or electronics departments within universities.

11. In all cases, it should be realized that the 'clean' gold surface will rapidly become contaminated, even upon brief exposure to ambient (atmospheric) conditions. The only truly clean environment is in the UHV. When possible, in order to reduce (but not negate) contamination, surfaces should therefore be stored/transported in ultrapure distilled water.

[a] Caution should be exercised, both in making up stock Pirahna solutions, and in their use. All operations should be carried out in a fume-hood and suitable protective clothing must be worn. This cleaning reagent reacts violently with organic materials. Extreme care should also be taken when washing away the acidic solution with water (see later). Finally it should also be noted that extended cleaning in Pirahna solution may not only increase the roughness of the gold films, but also may damage insulators (such as CVD silicon nitride) used for defining microstructures.

3.1 Deposition of alkanethiolate SAMs

The most frequently used technique to prepare organosulfur SAMs is solution self-assembly from either organic or aqueous solvents. The process [simply] involves the immersion of a clean, gold-coated substrate (*Protocol 1*) into a solution of the thiol, sulfide, or disulfide (typically at 1–2 mM concentrations). In the first instance, we focus on the use of the ω-substituted alkanethiols, since they have been most frequently used.

The properties of single component ω-substituted alkanthiolate SAMs on gold depends primarily on the temperature, the solvent used, the purity of the thiol, the immersion time, and on the quality of the gold substrate. The purity of the compound may at first glance be considered important when dealing with thiols since it is well-known that thiols readily dimerize to form a disulfide when exposed to oxygen. However, a small fraction of the corresponding disulfide in the solution is not critical for the formation of single component monolayers, because disulfides are known to undergo S–S bond reductive

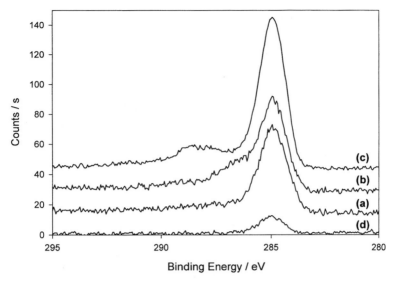

Figure 2. C(1s) XPS spectra of thin film gold electrodes evaporated on to glass substrates. Take-off angle 10°, flood gun kinetic energy 0 eV (binding energies corrected using Au (4f) spectra). (a) Sample measured within 4 h of evaporation. Note the low level of non C–C, or C–H species. (b) Sample as in (a) immersed in phosphate buffer for 1 h. Note the increase in the spectra at c. 286 eV corresponding to C–O species. (c) Sample measured two days after evaporation (stored within a closed container in the laboratory). Note the increase in spectra at c. 288 eV corresponding to C=O species (possibily human 'contamination' such as skin rather than hydrocarbon contamination, which has no C=C =O). (d) Sample from (a) after 2 min Ar⁺ etching, showing (only) residual carbon due to contamination within Au film.

cleavage and adsorb as thiolates on gold. The kinetics of SAM formation is, however, slower for disulfides, a feature which will influence the preparation of mixed SAMs (see below). The critical steps which improve the quality of the SAM are given in *Protocol 2*.

3.2 Preparation of alkanethiolate SAMs on gold

The vast majority of the alkanethiolate SAMs have been prepared from ultrapure ethanol solution. Ethanol is a good solvent for *n*-alkanethiols up to a chain length of about 18 methylene units ($n = 18$). Above 18 methylenes, the compounds display a tendency to precipitate, and more appropriate solvents for long chain alkanethiols ($n > 18$) are, e.g. hexane, dimethyl ether, or tetrahydrofuran. For shorter chain thiols, the assembly process may be performed from aqueous solution (e.g. with cysteine, mercapoethanol, and mercaptoproprionic acid).

Protocol 2. Preparation of single component alkanethiolate
SAMs on gold

Reagents
- Appropriate thiol compound
- Ethanol

Method

1. Recrystallize or vacuum distil the alkanethiol sample in order to ensure a low level of the corresponding disulfide in the loading solution.

2. Immediately prior to the experiment, dissolve the alkanethiol at a concentration of 2 mM in an appropriate solvent. It is recommended that the solvent be degassed in order to remove oxygen. Use clean glass or plastic (e.g. polypropylene or Teflon) containers.

3. Remove the freshly cleaned gold substrate from its storage solution (see *Protocol 1*), and immerse it in the thiol-containing solution for at least 12 h. Although the assembly process is very fast for *n*-alkanethiols on gold, the quality of the SAM in terms of chain orientation, organization, the molecular packing (density) improves substantially if the SAM is allowed to equilibrate with the solution for prolonged periods.

4. Remove the gold coated substrate(s) from the thiol solution and rinse exhaustively in pure solvent. In some cases it may also be necessary to remove physisorbed SAM material by placing it in a sonic bath for a maximum of 2 min.

5. Make duplicate samples so that they can be independently characterized by, e.g. ellipsometry, contact angle goniometry, cyclic voltammetry, infrared and X-ray photoelectron spectroscopy (see *Table 1*). Mount the substrate in the sensor or the instrument and use it immediately. Monolayer structures are sensitive to contamination (see *Figure 2*) (particularly prior to further functionalization). For example, rapid reorganization of the outermost portion of the tails has been observed for simple single component –OH and –COOH terminated SAMs.

6. Derivatize the SAMs with appropriate reagent, hapten, receptor, enzyme, antigen, or use it as prepared (see *Protocol 4 or 5*).

3.3 The preparation of mixed alkanethiolate SAMs on gold

The strong pinning of the sulfur to the substrate enables the preparation of mixed SAMs from ω-substituted alkanethiols, e.g. as $(HS–(CH_2)_n–X)f$ with $(HS–(CH_2)_n–Y)1–f$, $0 < f < 1$. By judicious choice of the components used in the self-assembly process, both the monolayer composition and the chain conformation can be controlled. Thus, it is possible to prepare surfaces with well-defined chemical compositions (X/Y functionalities), which also display a varying degree of crystallinity, as is schematically illustrated in *Figure 3* for an idealized mixed-model system of $HS-(CH_2)_{16}-OH)0.5/(HS-(CH_2)_n-CH_3)$ 0.5, where n is in the range 9–21. The understanding of the phase behaviour is, however, far from complete. It is known that phase segregation into macroscopic domains (micron sized) does not occur (10, 15), although phase segregation into microscopic aggregates has been observed (16).

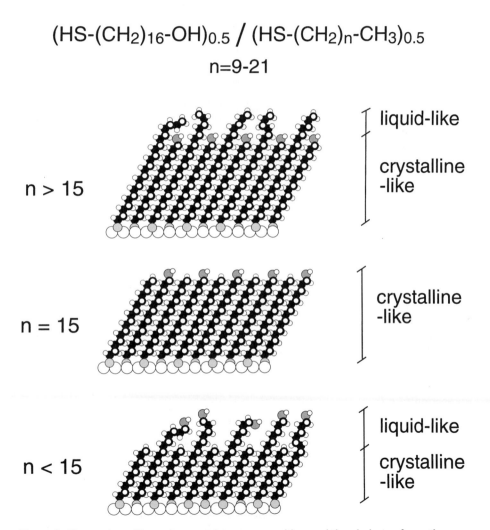

$$(HS-(CH_2)_{16}-OH)_{0.5} / (HS-(CH_2)_n-CH_3)_{0.5}$$
$$n=9-21$$

n > 15

liquid-like

crystalline -like

n = 15

crystalline -like

n < 15

liquid-like

crystalline -like

Figure 3. Illustration of how the monolayer composition and the chain conformation can result in a varying degree of crystallinity.

Density of receptor sites

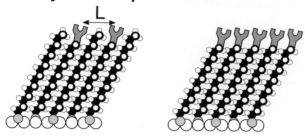

Mobility, accessibility, steric hindrance

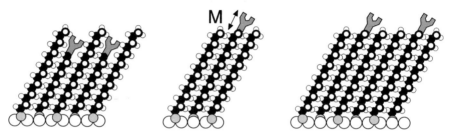

Figure 4. Illustration of how density of reconition sites at an interface are crucial parameters used during design of such interfaces, and can be used to manipulate the mobility, the accessibility and the degree of steric hindrance for a receptor (Y).

Mixed SAMs do not necessarily follow simple arithmetical rules of combination on all levels of molecular complexity. Care must therefore be taken when ω-substituted alkanethiols with different tail groups and chain lengths are combined to prepare mixed SAMs. Nevertheless, mixed monolayers of thiols on gold have been successfully used to optimize the properties of artificial sensing interfaces (3), and in *Figure 4* we identify the two most crucial parameters used during design of such interfaces, namely the density of recognition (receptor) sites and their mobility.

The recommended procedure for the preparation of mixed SAMs involves following the basic instructions given in *Protocol 2*, for single component SAMs. This protocol can be adapted as one of three methodologies (the first two being simpler than the third). The purity of the alkanethiols is more critical for the preparation of mixed SAMs, and care should be taken to ensure the purity of the starting materials (see *Protocol 2*). Large, uncontrolled, levels of the corresponding disulfides in the binary mixture will lead to an erroneous solution composition and hence to irreproducible surface compositions.

Protocol 3.   Preparation of mixed alkanethiolate SAMs on gold

Reagents

- Two appropriate thiol systems e.g. SH−
  $(CH_2)_{16}$−OH; SH−$(CH_2)_9$−CH_3
- Ethanol

A. *Mixed loading*

1. Follow *Protocols 1* and *2*, but add the gold substrate to a loading solution, comprising a mixture of two (or more) different stock solutions in appropriate molar ratios. Care should be taken to ensure that the solubilities of the thiols in the solvent used are comparable.

2. Ensure that the components in the loading solution are well mixed, if necessary by using ultrasonication. The final ratio of the components in the mixed SAM will be related to the concentrations of their components within the loading solution and the length of time of the exposure. The mixed monolayer can be readily characterized in detail using either FT-IR or XPS.

B. *Displacement*

1. Follow *Protocols 1* and *2*, adding the clean gold substrate to a loading solution comprising a single component stock solution.

2. Remove and wash the SAM in the pure solvent, as described in *Protocol 2*, and then place it in a solution of the second component SAM. The final ratio of the components in the mixed SAM will be related to the length of exposure to, and the concentration of, the second component SAM loading solution.

C. *Cross-diffusion*

1. It has recently been shown that it is possible to extend the complexity of mixed SAMs and prepare a chemical gradient on gold from ω-substituted alkanethiols by using a cross-diffusion process (17).

2. The clean gold substrate is exposed to two solutions of single component ω-substituted alkanethiols, from different directions. Establishing a capillary flow above the gold surface, using a clean glass slide and Teflon spacers facilitates the directionality of the cross-diffusion. Each stock solution is introduced simultaneously from opposite ends of the capillary. Further optimization of the conditions, for a given system, can be developed by drawing on the experience of others (17).

3.4 Biological self-assembly at mixed self-assembled monolayers for protein immobilization

The relevance of recognition, site density and mobility, see *Figure 4*, can best be understood through the examination of a bioanalytical interfacial system, first described by Knoll *et al.* (3), who examined mixed SAMs of a simple thiol alcohol and a complex biotinylated analogue, during the development of complex biosensing architectures based on streptavidin–biotin coupling (see *Figure 5*).

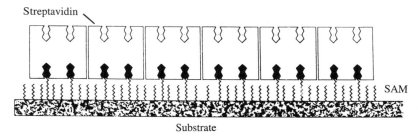

Dilution of Biotin-thiol with short chain spacer-thiol gives optimum packing of protein monolayer

Figure 5. A schematic of how the biotinylated thiol/thiol mixed monolayer can be used as a template to promote biological assembly of streptavidin. This process can be viewed as analogous to 'biological Lego'.

Protocol 4. Protocol for the immobilization of antibodies via a close-packed avidin/streptavidin monolayer on a modified gold surface

Reagents

- Streptavidin
- 11-mercaptoundecanol
- MBDD

- PBS
- Biotylated 1 gG

68

Method

The protocol enables the deposition of streptavidin onto a mixed mono-layer, created via the adsorption of the protein onto a mixed self-assembled monolayer which contains two monolayer forming alkanethiol derivatives: 11-mercaptoundecanol and 12-mercapto(8-biotinamide 3,6-dioxaoctyl) dodecanamide (MBDD), both of which can be obtained from Boehringer Mannheim GmbH. Having produced the streptavidin monolayer, biotinylated antibodies can be readily assembled onto the surface.

1. A clean, dry, gold surface is first derivatized with the mixed alkanethiol monolayer. This is achieved by the immersion of the substrate in an ethanolic (99.99+ purity) solution of the two alkanethiol derivatives for 1 h (see *Protocol 3*, part A). To ensure optimal binding of protein, the mole fraction of the biotinylated thiol in solution must be 0.1, i.e. a 9:1 ratio of the alcohol thiol to the biotinylated thiol. The total concentration of alkanethiol in solution should be between 0.5–1 mM.

2. After immersion, the substrate should be thoroughly rinsed, first in ethanol and then in ultrapure water. During rinsing the substrate should demonstrate hydrophilic behaviour—if this is not the case, then it is likely that incomplete monolayer transfer has occurred, and the sample should be discarded. After rinsing, the substrate should be dried in a flowing stream of nitrogen gas. If the substrate is not going to be used immediately, it should be stored in ultrapure water.

3. Protein adsorption is achieved by immersion of the freshly derivatized substrate in a 0.5–1 μM solution of avidin or streptavidin for 1–2 h. The streptavidin should be dissolved in 10 mM phosphate-buffered saline (PBS) pH 7.4.

4. On removal from the protein solution, the substrate should be thoroughly washed, first in PBS and then in ultrapure water. Again, during rinsing the substrate should exhibit hydrophilic behaviour. The streptavidin derivatized substrate is now ready for further derivatization with biotinylated molecules or for storage. *Figure 5* shows a schematic of the assembly process, which can be described as 'biological lego'. The maximum storage time for the biofunctionalized interface in PBS at 4°C is approximately 36 h.

5. Antibodies can now be readily immobilized through 'biological self-assembly', if the surface is simply immersed in a solution of biotinylated IgG (2 mg/ml in PBS pH 7.4), available from Sigma.

3.5 Immobilization of antibodies on gold

As an alternative to using the biotinylated thiol self-assembling system described above (*Protocol 4*), it is possible to covalently attach a biomolecule

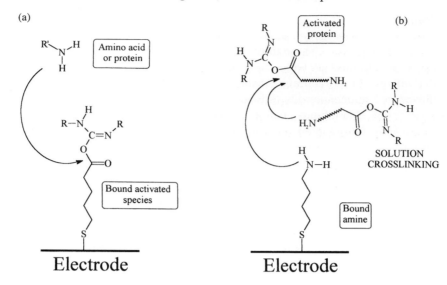

Figure 6. (a) Carbodiimide activation of a –COOH terminated SAM, which then undergoes nucleophilic substitution in the presence of the primary amine such as a surface exposed lysine in the protein structure. (b) It is feasible to use a complementary methodology to bind carbodiimide 'activated' proteins to a –NH$_2$ terminated SAM although this methodology can lead to cross-linking.

(such as a ligand-binding protein, i.e. either an antibody or (strept)avidin) onto a SAM-modified gold surface. Of the immobilization strategies available, it is recommended to use –COOH terminated groups ($16 > n \geqslant 3$) e.g. mercaptopropionic acid, $n = 3$). These SAMs can then be activated by using a carbodiimide reaction and will undergo nucleophilic substitution in the presence of the primary amine such as surface exposed lysine residues in the protein structure (*Figure 6a*). The hydrophilic nature of lysines ensures that these predominate on protein surfaces. Whilst, in principle, it is feasible to use a complementary methodology to bind 'activated' proteins to a –NH$_2$ terminated group (e.g. mercatoethylamine), in practice this protocol can lead to the protein–protein cross-linking (*Figure 6b*).

Protocol 5. Electrochemical and spectrophotometric 'sandwich' immunoassays with the antibody immobilized on gold surfaces

Reagents
- Ethanol
- α-Napthyl phosphate
- Mercaptopropionic acid

Method

Depending upon whether an amperometric or a spectrophotometric assay is to be used, two suitable gold surfaces (or in the case of the electrochemical assay 'electrodes') must be prepared. These can be either a pair of polycrystalline electrodes or two 'identical' evaporated gold substrates, bonded to wires with silver dag (quick drying silver paint, from Agar Scientific) and sealed with silicon sealant (Sealguard, Corning). One of the electrodes will be used for the experimental measurement, whilst the second will constitute the control experiment.

1. A pair of 'identical' gold surfaces are first cleaned (*Protocol 1*) and modified using a 2 mM solution of an appropriate thiol, e.g. mercatopropionic acid (*Protocol 2*), in a suitable solvent system, such as ethanol. The substrates are incubated for at least 12 h, but preferably 24 h in a solution of the thiol. In either case, after functionalization with the SAM, the surface is then washed exhaustively using ultrapure or deionized water.

2. The –COOH terminated thiol SAM is activated by incubation with a water soluble carbodiimide, e.g. by immersion in a *freshly prepared* solution of 10% (w/v) of 1-ethyl-3-(3-dimethylaminopropyl) carbodiimide (EDC) in 10 mM phosphate buffer pH 7.4 for 2 h.

3. The substrates are then washed using ultrapure water and are immersed in a solution of 10 μg/ml of a specific antibody in 10 mM PBS pH 7.4 for 1 h.

4. Both substrates are exposed to 10 mg/ml casein in 10 mM PBS pH 7.4, thus blocking any unmodified sites on the gold surface.

5. One gold substrate (the 'experiment') is incubated in 10 mg/ml antigen, whilst the other (the 'control') is incubated in 10 mg/ml of a control protein, both in 10 mM PBS pH 7.4 (both for 2 h).

6. Finally, both substrates are washed in PBS and are incubated in a solution of 10 μg/ml of an alkaline phosphatase–second antibody conjugate in 10 mM PBS pH 7.4 for 2 h.

7. After washing with PBS, the alkaline phosphatase activity can be assayed spectrophotometrically by placing the electrode in a plastic cuvette (path length 1 cm) and using 10 mM α-napthylphosphate as the substrate. Stir or agitate the solution during this period, and measure the change in the absorbance using a spectrophotometer ($\lambda = 404$ nm). Record a single 'end-point' absorbance reading after 5 min incubation.

8. Alternatively, the alkaline phosphatase activity can be assayed electrochemically (using chronoamperometry) at 320 mV versus Ag|AgCl. The baseline can be obtained in a 50 mM KCl solution (as the supporting electrolyte), whilst the measurement can be made using a 10 mM

Protocol 5. *Continued*

solution of α-napthylphosphate as the substrate (again, in a 50 mM KCl solution). Record an 'end-point' current reading after 5 min incubation.

9. The protocol can be repeated with different amounts of immobilized antibody (0.1–10 μg/ml) or different amounts of developing antibody (0.1–10 μg/ml), and suitable binding isotherms can be established.

10. In either case, whether measuring the reaction amperometrically or spectrophotometrically, the comparison between the two electrodes gives a direct comparison of a specific assay (measured using the antigen) and the nonspecific binding reaction (measured using the control protein).

4. Preparation of silane SAMs on silicon-based substrates

Having spent the first part of the chapter examining thiol deposition on Au, we will now turn our attention to the functionalization of silicon and glasses with silanes. Silanes are a group of molecules that consist of a silicon atom covalently attached to four variable groups. One or more of these groups can be readily substituted by the oxygen of a hydroxyl group at a silicon oxide surface. Thus, these molecules are capable of forming covalent bonds with a variety of glasses and silicon substrates. As stated, some of the most thoroughly studied and well-characterized SAMs in this category are those prepared from alkyltrichlorosilanes (or chlorosilanes), e.g. octadecyltrichlorosilane (OTS). Other silanes commonly used include the aminosilanes e.g. aminoethylaminopropyltrimethoxysilane (which is also known as trimethoxylsilylpropylethylene diamine).

In this section, two silanization procedures are described. The first involves the functionalization of silicon with 3-mercaptopropyltrimethoxysilane (MPTS), promoting the adhesion of (evaporated) noble metals to SiO_2 and glasses (*Protocol 7*). In addition, a method for the immobilization of antibodies (*Protocol 8*) is also detailed. Prior to this, again, it is important to consider aspects of cleaning the substrates.

4.1 The preparation of substrates

As is the case for metallic substrates, it is important to clean the silicon or glass, which can be done most effectively using a low power dry etch, such as an Ar or an oxygen plasma (both of which may cause some roughening of the surface). As an alternative method, it is possible to a wet etch the surface, e.g using hydrofluoric acid (HF). In this procedure, we describe a less vigorous, but effective method for cleaning surfaces using degreasing agents, such as

Decon. This does not require the use of hazardous chemicals or expensive equipment.

Protocol 6. Cleaning of Si-based substrates

Reagents
- Decon
- H_2SO_4
- H_2O_2
- iPA

Method

1. Wear polythene gloves when handling samples since latex rubber gloves tend to leave behind a small residue of latex particles, or talc, which requires further, extensive, cleaning to remove.

2. If the substrates will not be damaged, wipe them with soft tissue soaked in iPA (isopropyl alcohol) to remove finger prints and any bulk dirt.

3. Rinse the wiped substrate with iPA and blow it dry with a nitrogen gas stream.

4. Sonicate the substrates in Decon (diluted with an equal volume of purified water), or a similar 'degreasing' solvent for *c.* 10 min, and then rinse with copious amounts of ultrapure (distilled or deionized) water. Blow the substrate dry and bake in 105 °C oven for *c.* 10 min.

   Cleaning in hot Piranha solution can beneficially replace the above Decon clean, however, any perceived benefit of doing this must be off-set by the added care needed when handling such solutions, together with the risks of oxidative and/or thermal shock. Thermal shock to the material is particularly applicable when using large substrates (such as quartz blocks), as well as during dilution of the H_2SO_4 during the rinsing stages. Serial dilution of the Piranha solution using small quantities of water or more dilute sulfuric acid will avoid excessive heating.

4.2 Silanization of silicon or glass substrates

This section describes a procedure for coating of glass substrates with a mer-captosilane (3-mercaptopropyltrimethoxysilane, MPTS) prior to Au or Pt evaporation. Note that this modification can be used equally well for pro-moting the adhesion of silicon and glasses when deposited onto gold, as well as vice versa (see below). In addition it may be used for biological immobiliza-tions involving S–S linkages.

Suitable holders for containing the substrates in the reflux vessel can be made out of PTFE or glass. Metal and most plastics should be avoided (the former will react with the thiol groups, and the latter possibly dissolve leading to coating of the substrate with something other than the intended surface modifier).

Protocol 7. Deposition of silanes 'monolayers' on SiO_2 and glasses as adhesion layers to promote the attachment of gold and platinum

Reagents
- MPTS
- iPA

Method

1. Always use fresh MPTS (Sigma), as the cost of the silane is low relative to cost of the gold used in the evaporation. Silanes should be kept refrigerated prior to use, and allowed to come to room temperature before opening, thus avoiding condensation of atmospheric moisture into the stock solution. Old solutions oligomerize and the mercapto groups oxidize, thus making them less effective in bonding the evaporated Au.

2. Place the substrates (which may typically be several glass slides) in a suitable holder (slide rack) and immerse in a wide brimmed reflux vessel containing a solution of 400 ml iPA, 10 g MPTS, and 10 g H_2O.

3. Reflux the solution for *c*. 10 min and then place the reflux vessel in a bowl of cold water to speed cooling of the solution prior to the removal of the substrates.

4. On removal from the cooled solution, place the substrates in a beaker of iPA and leave for several minutes to allow the silane-containing solution to be rinsed off the substrate/holder surfaces.

5. Rinse well with iPA, blow dry as above, and then bake in 105°C oven for 10 min. If the oven temperature exceeds 115°C, the thiol will always oxidize and the gold will not adhere. Remember the coating is only a few atoms thick, and the oxidation reaction is irreversible.

6. The procedure for coating the substrate with the mercaptosilane should be repeated at least once, in order to ensure complete coverage.

7. Evaporate the gold or platinum (100 nm) within one day of preparing samples, otherwise the substrates will get covered in adventitious 'dirt' and an increasing proportion of the modifier's thiol groups will oxidize. Note in the authors' laboratories it has been shown using XPS measurements that after two days 50% of the surface sulfur species (which will interact with the evaporated metal) have been oxidized. See *Figure 2*.

The most likely cause of failure of gold adhesion when using this method is the oxidation or cleavage of the surface thiol species, which will occur if the temperature of the substrate increases greatly (i.e. either during the baking stage or during the evaporation of the metal). Of the two principle methods for gold evaporation, electron beam evaporation is preferred to thermal systems, as it has the advantage that generally the evaporator operates at lower base pressures with the consequence that the temperature of the molten metal is lowered for a given evaporation rate. The substrate temperature is influenced by both the temperature of the evaporated metal, the rate of metal deposition, and the radiant heat emitted from the evaporation crucible and surrounding areas of the source.

Radiant heat can be greatly influenced by the correct choice of evaporator system and its components: in electron beam systems, only a small region of the metal source is heated, and thus this has the lowest amounts of extraneous radiant heat. When using a thermal system, there is a choice between tungsten or molybdenum baskets/wires, or alumina coated, tungsten wire boats. Whilst alumina boats require outgassing before exposing the substrate to the molten metal, the power used in evaporating gold at a given evaporation rate is less than 25% that used when plain tungsten/molybdenum boats are used. This leads to a dramatic reduction in radiant heat and significantly improved gold adhesion (again these evaporation boats are more expensive, but are more durable and lead to greater success in fabrication procedures, and hence may be worth the additional expense).

4.3 The immobilization of biological molecules on silane layers

There are a wide range of options for the silanization of silicon-based sub-strates, and in this chapter, we have chosen to give one example, namely, using aminosilane. This is a complicated reaction which is carried out in the presence of water. Its three methoxy groups are reactive and can undergo three different reactions, see *Figure 7* and below.

(a) They either react with a hydroxyl group in the glass surface, establishing a covalent bond.

(b) Alternatively, they can react with a water molecule, so that a hydroxyl group is left in their place.

(c) Finally, they can react with the hydroxyl group of a silane molecule that has undergone the second reaction, establishing a cross-linking bond (5).

It follows from this simple description of the silanization process that the product of an extensive reaction of aminosilane with glass is not likely to be a monomolecular layer (as in the case of chlorosilane). Many of the silane molecules will be cross-linked, with the most 'conspicuous' groups exposed on the surface of the glass being the amino groups (see *Figure 7*). Glutaraldehyde

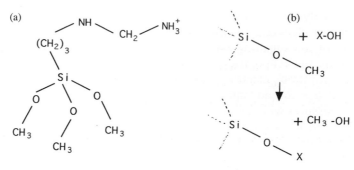

Figure 7. (a) A diagram of an aminosilane molecule is depicted (*left*). (b) The reaction of the aminosilane's methoxy groups is shown (*right*), where × can be the glass surface, hydrogen (i.e. water), or another aminosilane molecule. The role of water in determining the extent of cross-linking with other aminosilane molecules is apparent from this schematic.

(glutaric acid dialdehyde), which is a small linear organic molecule with an aldehyde group at each of its two ends, can subsequently be used as a cross-linker, to attach proteins to the amino groups (of the aminosilane) on the glass slides. The aldehyde groups will react with amino groups, either of the aminosilane or the protein to form Schiff bases, liberating a water molecule, thus anchoring the protein to the slide (18).

Schiff bases are generally unstable bonds, so it is preferable to stabilize them by reducing them with either sodium borohydride ($NaBH_4$) or with sodium cyanoborohydride ($NaCNBH_3$), in which case the bond would become an amine linkage. The convenience of using $NaCNBH_3$ instead of $NaBH_4$ can be best understood when considering the reductive methylation of the amino groups of a protein. Traditionally, this can be done by incubating the protein, first with formaldehyde and then with $NaBH_4$. Although the $NaBH_4$ reduces the Schiff bases to form secondary and tertiary amines, it also has the potential to reduce the aldehyde to produce an alcohol. The efficiency of protein immobilization is therefore a result of the competition between the two possible reductive pathways. On the other hand, as $NaCNBH_3$ readily reduces Schiff bases but not aldehyde groups, the efficiency of protein attachment is much higher. In addition, it can also be noted that whereas $NaBH_4$ has to be used at high pH (about 9), $NaCNBH_3$ can be used at neutral pH 7.

Protocol 8. Covalent binding of an antibody to a silanized SiO_2 or glass slide

Reagents

- Ethanol
- Glutaraldehyde
- Casein

- IgG-TRIC
- 1.3-Trimethoxysilylproplethylene
- Cyanoborohydride (sodium salt)

Method

1. The SiO_2 surface was silanized by incubation of a clean substrate in 2% 1,3 trimethoxylsilylpropylethylene diamine in 95% ethanol, 5% water for 10 min at room temperature. The wafer was rinsed in 95% ethanol, 5% water, and then was heated to 120°C for 30 min

2. The substrate is then immersed in 2% glutaraldehyde in 10 mM PBS pH 7.4.

3. Polyclonal antibodies (10 μg/ml) were immobilized to the silanized surface in a solution containing 40 mM sodium cyanoborohydride ($NaCNBH_3$) in 50 mM PBS pH 7.4 at 4°C for 16 h.

4. The surface was then exposed to 10 mg/ml casein in PBS pH 7.4, so blocking unreacted sites on the silicon surface.

5. The binding of the antibody was examined by performing a hetero-geneous sandwich 'immunospot' assay. This involved incubating the antibody functionalized surface in a solution of a 10 μg/ml tetramethyl-rhodamine isothiocyanate (TRITC) labelled protein antigen in 10 mM phosphate buffer pH 7.4.

6. The surface is washed in 10 mM phosphate buffer pH 7.4, and visual-ized using fluorescence microscopy (Nikon Microphot, Nikon UK Ltd.).

7. The amount of binding can be visualized by using a CCD camera (with a standard personal computer with a PC 'frame-grabbing' card). Images can subsequently be quantified into a suitable image process-ing package such as NIH image (available by ftp from `zippy.nih.gov`).

8. The amount of non-specific binding can be determined at different stages in this process. For example, the process can be repeated with-out 'blocking' the surface with caesin, hence showing an increased amount of binding due to NSB on the unblocked surface (which itself is a measure of the efficiency of the antibody binding to the silane).

5. Conclusion

In conclusion, this chapter introduces the reader to the bioanalytical applica-tions of self-assembled monolayers, concentrating on the practicalities of forming both thiol monolayer formation on gold surfaces, and silanization of glasses and silicon. Inevitably, we must re-emphasize that, as the systems that we describe are monolayers, there is very little material immobilized on the surface, and whilst this provides some prospect of creating 'order' at the sur-face, the challenge still remains to stabilize these thin layers, so that inevitable losses in biological activity do not effect the overall reproducibility of the device being studied.

References

1. Ulman, A. (1991). *An introduction to ultrathin organic films: from Langmuir-Blodgett to self-assembly*. Academic Press, San Diego, CA.
2. Ulman, A. (1996). *Chem. Rev.*, **96**, 1533.
3. Mittler-Neher, S., Spinke, J., Liley, M., Nelles, G., Weisser, M., Back, R., *et al.* (1995). *Biosensors Bioelectron.*, **10**, 903.
4. Parikh, A.N., Liedberg, B., Atre, S.V., Ho, M., and Allara, D.L. (1995). *J. Phys. Chem.*, **99**, 9996.
5. Vankan, J.M.J., Ponjee, J.J., Dehaan, J.W., and Vandeven, L.J.M. (1988). *J. Coll. Interface Sci.*, **126**, 604.
6. Nuzzo, R.G. and Allara, D.L. (1983). *J. Am. Chem. Soc.*, **105**, 4481.
7. Fujita, M., Yazaki, J., Kuramochi, T., and Ogura, K. (1993). *Bull. Chem. Soc. Jap.*, **66**, 1837.
8. Dubois, L.H. and Nuzzo, R.G. (1992). *Annu. Rev. Phys. Chem.*, **43**, 437.
9. Azzam, R.M. (1987). *Ellipsometry and polarized light*. North Holland Physics Publishing, Amsterdam, The Netherlands.
10. Bain, C.D. and Whitesides, G.M. (1989). *J. Am. Chem. Soc.*, **111**, 7164.
11. Laibinis, P.E., Nuzzo, R.G., and Whitesides, G.M. (1992). *J. Phys. Chem.*, **96**, 5097.
12. Porter, M.D., Bright, T.B., Allara, D.L., and Chidsey, C.E.D. (1987). *J. Am. Chem. Soc.*, **109**, 3559.
13. Yates Jr, J.T. and Madey, T.E. (ed.) (1987). *Vibrational spectroscopy of molecules on surfaces*. Plenum Press, New York.
14. Siegbahn, K., Nordling, C., Fahlman, A., Nordenberg, R., Hamrin, K., Hedman, J., *et al.* (1967). ESCA-atomic, molecular, and solid state structure studied by means of electron spectroscopy. *Nova Acta Rev. Sci. Ups.* Ser. IV, Vol. 20.
15. Bertilsson, L. and Liedberg, B. (1993). *Langmuir*, **9**, 141.
16. Stranick, S.J., Parikh, A.N., Tao, Y.-T., Allara, D.L., and Weiss, P. (1994). *J. Phys. Chem.*, **98**, 7636.
17. Liedberg, B. and Tengvall, P. (1995). *Langmuir*, **11**, 3821.
18. Ikariyama, Y. and Aizawa, M. (1988). In *Methods in enzymology* Vol. 137, p. 111.

5

Protein adsorption: friend or foe?

ERIKA JOHNSTON and BUDDY D. RATNER

1. Introduction

The phenomenon of protein adsorption is widely observed in nature and in technology. For example, consider marine fouling, contact lenses, food processing equipment, cell growth surfaces, and biosensors. Also, the biocompatibility of medical implants may be driven by the surface adsorbed layer of proteins.

When a biological solution comes in contact with a clean surface, proteins and other surface active molecules diffuse rapidly from the bulk liquid and become reversibly or irreversibly adsorbed to the surface (1). Although the compositional mixture, conformation, and binding strength of the adsorbed proteins may vary with time, the conditioning layer is persistent and will mediate all future interactions between the surface and components in the surrounding liquid. The effects attributable to conditioning layers can be separated into two types: those directly associated with the conditioning layer itself (changes in permeability, lubricity, texture, etc.) and those due to cells (e.g. macrophages, platelets, bacteria) that are activated or attracted by the adsorbed material.

Two broad materials approaches have been taken to address adsorbed proteins at the solid–liquid interface and their role in biology and medicine:

(a) Inhibit the proteins from attaching to the surface.

(b) Develop surfaces that direct the correct protein (ideally in the correct conformation) to the surface and retain it there.

This review will outline surface modification methodologies used to achieve both goals. Surfaces for covalent immobilization of proteins will not be covered in this article. Rather, it will focus on surfaces that spontaneously resist or retain high levels of protein after simple adsorption from aqueous solution. Surface characterization of these protein-resistant or protein-retentive surface modifications is essential to ensure accurate control of surface structure (and, hence, protein interaction) and for quality control. Surface characterization methods will not be reviewed here, but many articles are available overviewing these analytical methods (2–4).

2. Strategies to inhibit protein adsorption

2.1 General concepts

The detrimental effects of conditioning layers are illustrated by the case of chemical sensor fouling. Non-specifically adsorbed materials can line and occlude the pores of protective membranes used in many sensors. Recognition and transduction sites can be physically blocked or deactivated. Also, for sensors that detect the mass of specifically adsorbed molecules, e.g. standing acoustic wave (SAW) devices, or changes in refractive index at a solid–liquid interface, e.g. surface plasmon resonance (SPR) devices, non-specifically adsorbed material constitutes an interferent that is time variant and difficult to offset.

The effects of fouling by cellular attachment are specific to certain biological environments. At medical implant surfaces, adsorption of plasma proteins such as fibrinogen and fibronectin promote the adhesion and activation of platelets which can lead to inflammation, thrombus formation, and walling off of the device (5). In industrial and marine environments, bacterial and algal biofilms can clog desalination membranes, block optical windows (6), and add viscous drag to ship hulls (7) and piping systems (8). It has been hypothesized that the initial stages of biofilm formation are related to the presence of conditioning films on surfaces. Once a cell becomes attached, the conditioning layer could contribute to its proliferation by providing a convenient and concentrated carbon source.

Poly(ethylene oxide)-like (PEO-like) and phospholipid-like modified surfaces both exhibit resistance to protein adsorption and cellular attachment relative to unmodified control surfaces. Theories for the effectiveness of phospholipid-like layers in inhibiting protein adsorption have been advanced by a few groups (9–13). The non-fouling properties of PEO are attributed to a combination of qualities: steric repulsion by the mobile strands (14), caging of water molecules by the ethylene oxide repeat unit (15), the absence of sites that can participate in long-range attraction of charged groups or short-range attraction of hydrophobic groups (14, 16), and the inability of proteins to 'melt' structured water on PEO (17).

Section 2 describes methods for preparing PEO-like and phospholipid-like surfaces to lessen fouling in biological environments. It should be noted that individual applications will require optimization to produce stable surface modifications with good coverage and non-fouling properties. Three primary considerations drive the selection of a non-fouling surface modification strategy:

(a) The nature of the substrate to be modified (composition, porosity, geometry, etc.).

(b) The demands of the ultimate application (transparency, minimal diffusion barrier, type of biological challenge, etc.).

(c) The availability of organic synthesis expertise and specialized equipment.

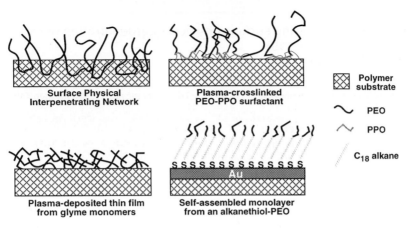

Figure 1. Methods of immobilizing PEO on surfaces to reduce protein and cell interaction.

As each technique is presented these considerations will be addressed. To clarify the differences in the methods, each of the PEO-based techniques for inhibiting protein pickup on surfaces is illustrated schematically in *Figure 1*.

2.2 Surface physical interpenetrating network (SPIN)

The SPIN technique, developed by J. Hubbell and colleagues, is attractive for its simplicity since the reagents are widely available and the procedure can be performed with standard laboratory equipment. This technique uses a solvent to interrupt the intermolecular bonds in a solid polymer sufficiently to enable PEO chains to diffuse into the surface and become intermixed with surface polymer chains. The diffusion is stopped abruptly by adding an excess of non-solvent, trapping the PEO strands in the polymer surface. SPIN modified surfaces demonstrate reduced protein adsorption and cellular attachment relative to untreated poly(ethylene terephthalate) (PET) controls (18–20).

To prepare a SPIN surface, it is necessary to identify solvents which dissolve the PEO and swell the substrate polymer, without hydrolysing or degrading either polymer. Many sources of information on the ability of solvents to swell or dissolve polymers are available (21). As seen in *Protocol 1* for modification of polyurethane, solvent systems can be optimized by adding water, if pure solvent dissolves the polymer substrate. Also, if the PEO chain is too large, it will penetrate the polymer too slowly. If it is too small, it may not be effectively trapped by the surface. For the polymer solvent systems described in *Protocol 1*, 18.5 kDa was determined to be a useful PEO molecular weight, with 10 kDa being too small and 100 kDa too large. Surfaces treated with 18.5 kDa PEO were also the most effective in resisting albumin and fibrinogen adsorption and fibroblast adhesion (19).

Protocol 1. Preparing surface physical interpenetrating networks of PEO (19)

Reagents

- Polyurethane (PU) (Pellethane®, Dow Chemical, 2363–80AE)
- Poly(methyl methacrylate) (PMMA) (medium M_r, Aldrich Chemical Co.)
- Poly(ethylene terepthalate) (PET) (Mylar, DuPont)
- Tetrahydrofuran (THF)

- Trifluoroacetic acid (TFAA) (Morton Thiokol, Inc.)
- Poly(ethylene oxide) (PEO) (18.5 kDa, Polysciences, Inc.)
- Stock PEO solution: 40 g PEO in 100 ml deionized filtered water

A. *Method for PET*

1. Extract PET for at least 24 h in acetone at room temperature.

2. Add one part stock PEO solution to four parts TFAA to achieve an 8% (w/v) solution.

3. Place the PET sample in the PEO:TFAA solution and let it stand for 30 min at room temperature. Periodically swirl the solution.

4. Quench the reaction by rapidly adding a large excess of water.

5. Transfer the treated PET samples to deionized water for storage.

B. *Method for PU*

1. Dissolve Pellethane® pellets to a concentration of 50 g/litre in THF.

2. Cast film to a wet thickness of $\sim$ 1 mm and cure in an oven at 60–70°C. Cut to size.

3. Add 20 ml stock PEO solution to a flask containing 40 ml THF and 40 ml filtered deionized water.

4. Heat the solution to 60°C in an oven until the PEO dissolves completely.

5. Immerse Pellethane films in the PEO:THF solution for 15–25 min at 60°C. Swirl periodically.

6. Quench the reaction with an excess of water and transfer the films to a fresh change of water for storage.

C. *Method for PMMA*

1. Prepare samples by dissolving PMMA powder in acetone to a concentration of 50 g/litre.

2. Cast a PMMA film to a wet thickness of $\sim$ 1 mm and cure in an oven at 60–70°C.

3. Add 20 ml stock PEO solution to a flask containing 60 ml acetone and 20 ml deionized water.

4. Immerse PMMA films in solution for 15–25 min at room temperature. Periodically swirl the solution.
5. Quench the reaction with an excess of water and store the samples in a fresh change of water.

2.3 Grafting of adsorbed molecules using RF plasmas

Physisorbed layers of PEO are unstable in aqueous biological environments due to the high water solubility of PEO. Sheu and Hoffman were able to stabilize surface deposits of PEO surfactants by radiofrequency glow discharge grafting (22). Their studies suggest that vacuum UV radiation from an argon radio frequency (RF) plasma covalently cross-links Brij 99 and bonds it to an underlying polyethylene subtrate. The grafted chains were stable to soaking in chloroform and water and they reduced fibrinogen adsorption by 90% relative to untreated polyethylene.

The plasma reactor used in Sheu's study consisted of a glass cylindrical reactor 11.5 cm i.d. × 90 cm long. A 13.56 MHz RF generator excited the plasma via a matching network and a pair of capacitively coupled external copper electrodes. The electrodes were spaced 56 cm apart and the samples were placed half-way between them on the floor of the reactor. Flow of argon gas was controlled with a mass flow controller and a liquid nitrogen cold trap on the reactor outlet prevented traces of pump oil from backstreaming into the reactor. A reactor of this general type is illustrated in *Figure 2*.

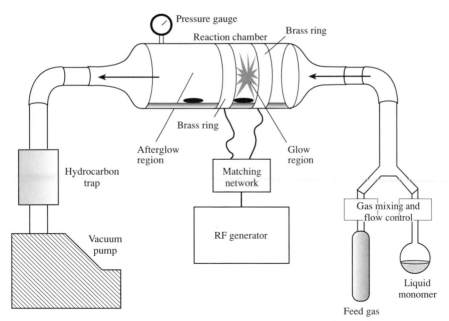

Figure 2. RF plasma reactor for surface cross-linking, etching, or deposition.

Prior to assembling the reactor, the glass chamber is baked out by ramping a glassware oven to 400 °C and cooling overnight. After assembling the reactor, the system should be cleaned once more by etching with a two standard cubic centimetre per second (sccm) stream of argon for 10 min at 50 W and 50 mtorr. Before loading samples, the glow discharge should be tuned to the desired power and pressure so that a stable glow can be achieved immediately after the plasma is excited.

Etch duration, plasma pressure, and plasma power all influenced the physical and chemical properties of the resulting film. Using 2.5 W power and 25 mtorr produced films which significantly retained PEO structure (i.e. relatively low oxidative degradation). Under these conditions, a 30–60 second etch time resulted in films with the best non-fouling properties and appropriate O:C ratios.

The PEO is introduced to the surface prior to plasma binding by dipping the substrate polymer in a PEO-containing surfactant solution. According to Sheu, it is best to choose a dipping solution that wets but does not otherwise degrade the substrate. From the thin liquid film that clings to the sample surface, a physical deposit of PEO or PEO surfactant can form when the solvent evaporates. If a thin liquid film is not retained on the sample surface, much less polymer is found to be covalently deposited after plasma treatment.

Protocol 2. Grafting of PEO and PEO-containing surfactants using RF plasmas (22)

Equipment and reagents
- Plasma reactor (see text)
- Low density polyethylene (LDPE), 0.3 mm thick (Cadillac Plastics)
- Brij 99 (Sigma Chemical Co.)
- Chloroform (J. T. Baker)

Method

1. Clean the sample surface—sonicate LDPE in methylene chloride, acetone, and water for 30 min each to remove possible plasticizers and additives.

2. Prepare a 1% (w/v) polymer solution of Brij 99 in chloroform.

3. Dip the sample in the polymer solution for 30 sec at room temperature.

4. Dry the sample in laminar flow-hood overnight to evaporate chloroform.

5. Assemble the reactor, and check for leaks by pumping down to a base pressure of ~ 10 mtorr. Ensure that the reverse power is minimized at the desired Ar pressure when the RF power is switched on, so that no further optimization is required.

6. Load the samples into the reactor.

7. Purge the reactor twice, pumping to < 15 mtorr and purging with argon to slightly above atmospheric pressure.

---

8. Pump down to < 10 mtorr for 15 min to remove any possible impurities remaining in the reactor. Purge again with argon and pump down. Introduce argon at a constant flow of 0.5 sccm. Manually shut off valves up- and downstream of the reactor when the desired pressure is reached.

9. Turn on the plasma to the pre-set forward and reverse power levels. Etch the samples for 30 sec and shut the power off. Bring the reactor up to atmospheric pressure with argon.

10. Wash the samples twice for 30 min in chloroform and soak the samples overnight in chloroform to remove loosely bound surfactant.

11. Remove the samples, dry them in a fume-hood, and store them in air.

2.4 Plasma deposited PEO-like films

The apparatus for plasma deposition of thin, overlayer films shares many features with the one described in the previous section for surfactant grafting. A schematic of the reactor is shown in *Figure 2*. A low molecular weight precursor is warmed to raise its vapour pressure and an internally heated mass flow controller feeds the vapour to an evacuated plasma reactor. During plasma excitation, a fraction of the precursor molecules enter excited states capable of binding to surfaces and a film gradually forms on the samples in the reactor. Our work on non-fouling materials has focused on films deposited from oligo(ethylene glycol) dimethyl ethers (the oligoglymes) and cyclic molecules containing the $(CH_2-CH_2-O)_n$ repeat unit (dioxane, 12 crown-4 ether and 15 crown-5 ether). Of these, the tri- and tetraglymes and the crown ethers reduce fibrinogen adsorption by 42–95% relative to glass and 81–98% relative to PTFE (23). The tri- and tetraglymes also reduce the attachment rate of bacteria relative to glass under parallel laminar flow (24).

Protocol 3 recommends a discharge profile that begins at high power and drops to a lower power after two minutes (23). This procedure forms a tightly adhesive, highly cross-linked underlying film that resists delamination from glass, polymers, gold films, sapphire, and porous membranes when exposed to water. The outer surface retains much of the ether carbon functionality that would be expected from a PEO-like surface.

Protocol 3.   Plasma deposited PEO-like films (23)

Equipment and reagents

- Plasma reactor with brass capacitor rings 15 cm apart (see *Figure 2*)
- Precursor: triethylene glycol dimethyl ether (triglyme) (Aldrich Chemical Co.)
- Round-bottom flask fitted with a thermocouple well
- Argon gas (99.8% pure)

Protocol 3. *Continued*

Method

1. Clean and dry the samples.
2. Pour 15 ml of triglyme precursor into a round-bottom flask and attach the flask to the internally heated mass flow controller.
3. Evacuate the reactor to a base pressure of 10–20 mtorr.
4. Freeze the triglyme in a liquid nitrogen bath and open the controller valve to evacuate the headspace of the flask. Check that the base pressure is still achievable.
5. Degas the triglyme by thawing it with a heat gun. Bubbles should appear as the triglyme melts. Freeze and thaw the precursor under vacuum at least three times until no gas bubbles can be seen escaping during the thaw.
6. Warm the precursor flask to ~ 94°C with heat tapes to raise the triglyme vapour pressure. Turn on the vapour feed to the reactor and ensure that the desired flow rate can be sustained for several minutes.
7. Shut off the triglyme vapour flow and evacuate the reactor. Etch the reactor with argon at 4 sccm for 30 min at 80 W.
8. Place the samples between the brass ring electrodes and evacuate the system to base pressure.

   Safety note: Take care not to touch the electrodes after power is activated.
9. Etch the samples with argon for 10 min at 80 W and 4 sccm. Evacuate the reactor to base pressure.
10. Deposit the film by turning on the vapour flow until the reactor pressure reaches 200 mtorr. Excite the plasma and tune it to 80 W power. Control the pressure to 250 mtorr. There should be a purple glow between the electrodes. After 2 min reduce the plasma power to 5 W for five more minutes. The glow should be barely visible.
11. After extinguishing the plasma glow, allow argon to flow over the samples for 10 min at 4 sccm. If condensate is visible on the walls of the reactor, evaporate it by heating locally with a heat gun or heat tapes. Evacuate the reactor to base pressure. Bleed argon into the reactor to bring it to atmospheric pressure and remove the samples.
12. Dip each sample four times in deionized water and then soak them in a fresh change of water for 10 h to remove monomer condensate or loosely bound plasma polymer.
13. Remove the samples from the soak solution and dry them for storage.

2.5 Self-assembled monolayers

Self-assembly of oligo(ethylene glycol) terminated thiolates is an attractive means for conferring PEO-like functionality to gold surfaces. This method is relatively simple and has potential for patterning (25) and incorporating recognition elements. Mixed monolayers of alkanethiolates and oligo(ethylene oxide) terminated alkanethiolates (EO-thiolate) have been systematically studied and shown to resist adsorption of a variety of proteins (26). The minimum fraction of oligo(ethylene oxide) alkanethiolates required to prevent protein adsorption decreased as the ethylene oxide chain length increased, and for thiolates containing 17 repeat units, fractional EO-thiolate coverages of 18–25% were sufficient. Drawbacks of the technique include its limitation to gold coated or other noble metal surfaces, the need for expensive and difficult to synthesize thiols, meticulous care in the preparation of the metal surface, and unproven long-term stability of the film.

Protocol 4.   Self-assembled monolayers of oligo(ethylene oxide) terminated thiols[a]

Equipment and reagents

- Metal evaporation apparatus
- Gold, chromium, and/or titanium evaporation targets
- Silicon wafers
- Oligo(ethyleneoxide) thiolate, n = 2–17[a]

Method

1. Prepare 0.25 mM solution of thiol in ethanol.

2. Evaporate gold onto an acid cleaned section of silicon wafer. As gold does not adhere well to silicon, first prime the silicon surface with a thin film of chromium or titanium.

3. Within an hour after gold evaporation, immerse gold coated sample in the solution. Reported immersion times range from minutes to 12 h (26). The films may be stored in thiol solution.[b]

4. Remove the samples, rinse them thoroughly with ethanol, and dry them under a stream of nitrogen.

[a] The procedure for synthesizing oligo(ethylene oxide) terminated thiols is available in supplementary material accompanying the original source (26).
[b] According to Prime (26), no change in composition of the films was observed over a period of several weeks.

2.6 Phospholipid layers

Phospholipid-containing polymers are hypothesized to adsorb and assemble free lipids from the blood, thereby presenting a biomimetic membrane surface to approaching proteins and platelets (10, 11). The hypothesis is supported by studies that show that as the phosphoryl choline component of the copolymer increases, the surface uptake of lipids from human plasma increases (12) and protein uptake decreases (11). Furthermore, platelet attachment and spreading from whole blood also decreased (12). The fact that compressed lipid layers in Langmuir troughs readily adsorb protein films is not well explained by the Ishihara hypothesis. Also, based on energetic considerations, the choline moeity, rather than the hydrocarbon chains will be oriented toward the aqueous phase. An alternative hypothesis, therefore, might attribute the low protein pickup of these materials to the strong water-binding of the choline group.

Preparation of a phosphotidyl choline surface requires synthesis of the copolymer MPC-co-BMA, which is a block copolymer of 2-methacryloyl oxyethyl phosphoryl choline (MPC) and butyl methacrylic acid (BMA). Procedures for synthesizing MPC are described in the original source (9).

Protocol 5.   Deposition of phospholipid-containing
  copolymers (11)

Reagents

- 2 methacryloyl oxyethyl phosphoryl choline (MPC) (9)
- 2,2'-azoisobutyronitrile (AIBN)
- Butyl methacrylate (BMA) (Fujikura Co. Ltd.)

Method

1. Polymerize MPC and BMA in ethanol at 60 °C for 15 h using AIBN as an initiator.
2. Dissolve poly(MPC-co-BMA) in ethanol (1 wt%).
3. Immerse the sample in poly(MPC-co-BMA) solution and remove.
4. Dry the sample and store in air.

---

3. Directing protein adsorption and retaining protein on a surface

3.1 General concepts

Interfacial proteins comprise the recognition and control element of the phenomenon we call life. Living cells respond with recognition and specificity to interfacial proteins, polysaccharides, and lipids. In order to precisely regulate cell attachment, spreading, differentiation, activation, and growth in biotechnology and medical devices, protein surfaces of synthetic systems must be engineered with precision. Also, other soluble biomolecules dock and interact with interfacial proteins to control processes in both *in vivo* and *in vitro* settings. *In vitro* applications where robust attachment of proteins is important include enzyme-linked immunosorbent assays (ELISA), biosensors, protein blotting protocols, affinity chromatography, and modern antibody diagnostic systems. Control of proteins adsorbed or bound at interfaces implies the following:

(a) Control of conformation (generally, but not always, retention of conformation is desired).

(b) Control of orientation (active site up).

(c) Control of which protein(s) are fractionated to the surface from the mixture of hundreds of proteins that make up most biological fluids (ideally, one specific protein is desired).

(d) Retention of protein to make a durable, robust system or device.

(e) Stability of the protein on the surface to time-dependent conformation rearrangement or proteolysis.

(f) Inhibition of non-specific protein adsorption (see Section 2).

A number of approaches have been used to retain and control proteins at interfaces. These include covalent immobilization to surface functional groups (27), assembly of proteins at surfaces (28), and immobilization to oriented lipid films (29–31). The simplest of the methods is to adsorb a protein to a surface in order to immobilize it. Proteins adsorbed to surfaces are often irreversibly bound, and they can retain considerable biological activity. Some relatively simple surfaces such as nitrocellulose and poly(vinylidene fluoride) have shown good ability to retain adsorbed protein. More sophisticated surfaces for protein retention can be fabricated using glow discharge or corona ionized gas processes. Commercial cell cultureware and dishes for ELISA assays fall under this category. Unfortunately, the protocol for manufacturing these surfaces is largely proprietary and unpublished. However, some surfaces that show excellent retention of protein and also unique biological interactions can be fabricated using published ionized gas surface treatment methods. The methodologies for the preparation of a few of these surfaces will be reviewed here.

3.2 Fluoropolymer films deposited from glow discharge plasma environments

Surfaces formed from glow discharge plasmas of fluorocarbon vapours (e.g. tetrafluoroethylene, hexafluoropropylene) exhibit unique interactions with proteins. Specifically, extremely tight binding of protein to the surface has been measured. Furthermore, plasma-deposited fluoropolymer surfaces demonstrate low blood platelet reactivity (32) and low adhesion to corneal endothelial cells (33).

Protein retention to surfaces can be measured by elution experiments. The use of a radiolabelled protein greatly simplifies this experiment. The protein is adsorbed to a surface. After the protein resides on the surface for a period of time, a surfactant such as sodium dodecyl sulfate (SDS) is used to wash protein from the surface. The fraction of protein that remains on the surface is measured. Typically, the longer the protein resides on the surface, the lower its elutability (higher retention). Also, elutability varies with protein type and with the nature of the surface. The elutability experiment has been described in a number of publications (34–36, 38, 39). For plasma-deposited fluoropolymer surfaces, after 2 h protein residence time, protein retentions higher than 90% are often measured. In contrast, polytetrafluoroethylene shows protein retentions typically around 50%.

The chemical structure of these plasma-deposited fluoropolymer thin films is complex. Using the surface analytical method, electron spectroscopy for chemical analysis (ESCA), we learn that these films consist of highly cross-linked chains that are rich in –CF$_3$, –CF$_2$, and –CF groups (37, 38). Films prepared in the afterglow of the plasma reactor (downstream from the glow region) have been found to have a somewhat simpler chemistry dominated by –CF$_2$ groups and superior retention of protein (38, 39).

Protocol 6. Immobilization of proteins to fluoropolymer surfaces via tight binding (38)

Equipment and reagents
- RF plasma reactor (see *Figure 2*)
- Hexafluoropropylene gas
- Argon gas (99.98% purity)
- Poly(ethylene terephthalate) (PET) cover-slips (Lux Thermanox, Nunc, Inc.)

Method
1. Connect a lecture bottle of C$_2$F$_6$ gas to a plasma reactor similar to that illustrated in *Figure 2* and described in Section 2.3.
2. Evacuate the reactor to base pressure. Etch the reactor with argon at 4 sccm for 30 min at 80 W.
3. Poly(ethylene terephthalate) (PET) 0.6 × 0.6 mm^2 substrates are ultrasonically cleaned sequentially in methylene chloride, acetone, and deionized water, and then dried.
4. Place the PET samples approx. 20 cm downstream from the electrodes and evacuate the system to base pressure.
5. Etch the samples with argon for 10 min at 80 W and 4 sccm. Evacuate the reactor to base pressure.
6. Bring the reactor to 200 mtorr pressure with C$_2$F$_6$ gas at 5 sccm and expose the samples to the plasma at 10 W power for 120 min.
7. After turning off the glow, allow monomer gas to flow over the samples for an additional 1 h.

8. Bring the reactor to atmospheric pressure with argon gas.

9. Films made in this way, when exposed to protein solutions, will rapidly adsorb protein and have high surface retentions of protein (> 90%), even when exposed to SDS solution.

3.3 Allylamine films deposited from glow discharge plasma environments

Allylamine (CH_2=CH–CH_2–NH_2) coated surfaces, formed from vapour using an RF plasma discharge, show unique interactions with adsorbed proteins. Where the adsorption amount of fibrinogen from plasma to most surfaces goes through a maximum as a function of plasma dilution (a phenomenon referred to as the Vroman effect), to allylamine coated surfaces, fibrinogen adsorption increases with increasing plasma concentration (40). Furthermore, both protein elutability and baboon platelet adhesion rapidly decreased with residence time of adsorbed plasma proteins on allylamine surfaces, suggestive that the allylamine surface is effective in altering the conformation of the adsorbed protein.

Protocol 7. Surfaces coated with allylamine using RF plasma discharge (40)

Equipment and reagents
- RF plasma reactor with brass capacitor rings 30 cm apart (*Figure 2*)
- Allylamine (spectral grade, Aldrich Chemical Company)
- Poly(ethylene terephthalate) (PET) coverslips (Lux Thermanox, Nunc, Inc.)
- Argon gas (99.98% purity)

Method

1. Connect a flask of degassed allylamine liquid to a plasma reactor similar to that illustrated in *Figure 2* and described in Section 2.3.

2. Evacuate the reactor to base pressure. Etch the reactor with argon at 4 sccm for 30 min at 80 W.

3. Poly(ethylene terephthalate) (PET) 0.6 × 0.6 mm² substrates are ultrasonically cleaned sequentially in methylene chloride, acetone, and deionized water, and then dried.

4. Place the PET samples centred within a 25 cm zone between the brass capacitor rings and evacuate the system to base pressure.

5. Etch the samples with argon for 5 min at 50 W and 250 mtorr. Evacuate the reactor to base pressure.

6. Bring the reactor to 200 mtorr pressure with allylamine vapour (obtained by warming the allylamine liquid) and expose the samples to the plasma at 30 W power for 10 min.

Protocol 7. *Continued*

7. After turning off the glow, allow monomer to flow over the samples for an additional 30 min.

8. Bring the reactor to atmospheric pressure with argon gas.

9. Films made in this way should be used within one week of preparation.

3.4 Glow discharge plasma deposited films from acetone, methanol, and formic acid

RF plasma deposited surfaces from vapours of low molecular weight organic liquids containing carbon and oxygen can make surfaces that tightly bind proteins (41). These surfaces also serve as excellent substrates for growing cells. Tighter binding of protein to the surface is achieved by increasing the oxygen content (as measured by ESCA) of the deposited film. Oxygen content can be increased by blending oxygen gas into the organic vapour flowing in to the reactor, or by using more highly oxygenated liquids as monomers.

Protocol 8. RF plasma deposited surfaces from carbon/oxygen organic liquids (42, 43)

Equipment and reagents

- RF plasma reactor with brass capacitor rings 30 cm apart (*Figure 2*)
- Acetone (J. T. Baker Chemical Company)
- Methanol (J. T. Baker Chemical Company)
- Argon gas (99.98% purity)

- Formic acid (J. T. Baker Chemical Company)
- Poly(ethylene terephthalate) (PET) coverslips (Lux Thermanox, Nunc, Inc.)
- Oxygen gas

Method

1. Connect a flask of the organic liquid to a plasma reactor similar to that illustrated in *Figure 2* and described in Section 2.3. Also connect an oxygen tank to the plasma reactor.

2. Evacuate the reactor. Etch the reactor with argon at 175 mtorr for 60 min at 100 W.

3. Poly(ethylene terephthalate) (PET) 0.6×0.6 mm^2 substrates are ultrasonically cleaned sequentially in methylene chloride, acetone, and deionized water, and then dried.

4. Place the PET samples within a 25 cm zone centred between the brass capacitor rings and evacuate the system to base pressure.

5. Etch the samples with argon for 5 min at 40 W and 175 mtorr. Evacuate the reactor to base pressure.

6. Bring the reactor to 200 mtorr pressure with one of the organic liquid vapours (obtained by warming the liquid) and expose the samples to

the plasma at 5 W power, 20–60 mtorr pressure, for 10 min with a total flow rate of 1 sccm. Alternately, using a mass flow controller, blend oxygen into the organic vapour. Up to 60% oxygen can be blended with the monomer.

7. After turning off the glow, allow monomer to flow over the samples for an additional 10 min.
8. Bring the reactor to atmospheric pressure with air.
9. Films made in this way should be used within five weeks of preparation.

Acknowledgements

Funding for some of the protocols described in this article, and for the preparation of this manuscript has been received from NIH grants HL19419, HL25951, and RR01296, NSF Engineering Research Center Grant EEC9529161, and from the Center for Process Analytical Chemistry, University of Washington.

References

1. Horbett, T.A. and Brash, J.L. (1995). *Proteins at interfaces II: fundamentals and applications.* American Chemical Society, Washington DC.
2. Perry, S.S. and Somorjai, G.A. (1994). *Anal. Chem.*, **66**, 403A.
3. Ratner, B.D. (1988). In *Surface characterization of biomaterials* (ed. B.D. Ratner), p. 13. Elsevier Science, Amsterdam.
4. Johnston, E.E. and Ratner, B.D. (1996). *J. Elect. Spectrosc. Relat. Phenom.*, **81**, 303.
5. Gristina, A.G. (1987). *Science*, **237**, 1588.
6. Carr-Brion, K.G., Dowdeswell, R.M., and Sanderson, M.L. (1994). *Process Control Qual.*, **5**, 267.
7. Loeb, G.I. (1981). *Naval Research Laboratory memorandum report no. 4412.* Naval Research Laboratory, Washington DC.
8. Characklis, W.G. and Escher, A.R. (1988). In *Marine biodeterioration: advanced techniques applicable to the Indian Ocean* (ed. E.M.-F. Thompson). Oxford & IBH Pub. Co., New Delhi.
9. Ishihara, K., Ueda, T., and Nakabayashi, N. (1990). *Polymer J.*, **22**, 355.
10. Ishihara, K., Tsuji, T., Kurosaki, T., and Nakabayashi, N. (1994). *J. Biomed. Mater. Res.*, **28**, 225.
11. Ishihara, K., Ziats, N.P., Tierney, B.P., Nakabayashi, N., and Anderson, J.M. (1991). *J. Biomed. Mater. Res.*, **25**, 1397.
12. Ishihara, K., Oshida, H., Endo, Y., Ueda, T., Watanabe, A., and Nakabayashi, N. (1992). *J. Biomed. Mater. Res.*, **26**, 1543.
13. Hayward, J.A., Johnston, D.S., and Chapman, D. (1985). *Ann. N. Y. Acad. Sci.*, **446**, 267.
14. Jeon, S.I., Lee, J.H., Andrade, J.D., and Gennes, P.G.D. (1990). *J. Coll. Interf. Sci.*, **142**, 149.

15. Arakawa, T. and Timasheff, S.N. (1985). *Biochemistry*, **24**, 6756.
16. Merrill, E.W. (1987). *Ann. N. Y. Acad. Sci.*, **516**, 196.
17. Ter-Minassian-Saraga, L. (1981). *J. Coll. Interf. Sci.*, **80**, 393.
18. Desai, N.P., Hossainy, S.F.A., and Hubbell, J.A. (1992). *Biomaterials*, **13**, 417.
19. Desai, N.P. and Hubbell, J.A. (1991). *Biomaterials*, **12**, 144.
20. Desai, N.P. and Hubbell, J.A. (1991). *J. Biomed. Mater. Res.*, **25**, 829.
21. Brandrup, J. and Immergut, E.H. (ed.) (1989). *Polymer handbook*, 3rd edn. Wiley, New York.
22. Sheu, M.-S. (1992). Ph. D. dissertation. University of Washington.
23. Johnston, E.E. (1997). Ph. D. dissertation. University of Washington.
24. Johnston, E.E., Ratner, B.D., and Bryers, J.D. (1997). In *Plasma treatments and deposition of polymers* (ed. R. d'Agostino). NATO ASI Series, Kluwer Academic Publishers, Dordrecht, The Netherlands.
25. Singhvi, R., Kumar, A., López, G.P., Stephanopoulos, G.N., Wang, D.I.C., and Whitesides, G.M. (1994). *Science*, **264**, 696.
26. Prime, K.L. and Whitesides, G.M. (1993). *J. Am. Chem. Soc.*, **115**, 10714.
27. Amador, S.M., Pachence, J.M., Fischetti, R., McCauley, J.P., Smith, A.B., and Blasie, J.K. (1993). *Langmuir*, **9**, 812.
28. Haggerty, L., Watson, B.A., Barteau, M.A., and Lenhoff, A.M. (1991). *J. Vac. Sci. Technol. B*, **9**, 1219.
29. Losche, M., Erdelen, C., Rump, E., Ringsdorf, H., Kjaer, K., and Vaknin, D. (1994). *Thin Solid Films*, **242**, 112.
30. Pachence, J.M., Amador, S., Maniara, G., Vanderkooi, J., Dutton, P.L., and Blasie, J.K. (1990). *Biophys. J.*, **58**, 379.
31. Pum, D. and Sleytr, U.B. (1994). *Thin Solid Films*, **244**, 882.
32. Kiaei, D., Hoffman, A.S., and Hanson, S.R. (1992). *J. Biomed. Mater. Res.*, **26**, 357.
33. Mateo, N.B. and Ratner, B.D. (1989). *Invest. Ophthalmol. Vis. Sci.*, **30**, 853.
34. Rapoza, R. J. and Horbett, T.A. (1990). *J. Coll. Interf. Sci.*, **136**, 480.
35. Bohnert, J.A. and Horbett, T.A. (1986). *J. Coll. Interf. Sci.*, **111**, 363.
36. Chinn, J.A., Posso, S.E., Horbett, T.A., and Ratner, B.D. (1991). *J. Biomed. Mater. Res.*, **25**, 535.
37. Castner, D.G., Lewis, K.B., Fischer, D.A., Ratner, B.D., and Gland, J.L. (1993). *Langmuir*, **9**, 537.
38. Favia, P., Perez-Luna, V.H., Boland, T., Castner, D.G., and Ratner, B.D. (1996). *Plasmas Polymers*, **1 (4)**, 299.
39. Kiaei, D., Hoffman, A.S., Horbett, T.A., and Lew, K.R. (1995). *J. Biomed. Mater. Res.*, **29**, 729.
40. Chinn, J.A., Ratner, B.D., and Horbett, T.A. (1992). *Biomaterials*, **13**, 322.
41. Ertel, S.I., Ratner, B.D., and Horbett, T.A. (1991). *J. Coll. Interf. Sci.*, **147**, 433.
42. Ertel, S., Ratner, B.D., and Horbett, T. (1990). *J. Biomed. Mater. Res.*, **24**, 1637.
43. Ertel, S.I., Ratner, B.D., and Horbett, T.A. (1991). *J. Biomater. Sci.*, **3**, 163.

6

Micropatterning cell adhesiveness

PETER CLARK

1. Introduction

The ability to micropattern adhesiveness for cells has been exploited by a number of research groups to provide model surfaces for investigating cell behaviour, and for devising technologies in cell and tissue engineering (1–11). These techniques have been developed to make surfaces which provide differentially adhesive cues for cultured cells, i.e. areas which are permissive for cell adhesion and locomotion, with adjacent non-permissive areas. The techniques were developed using a variety of approaches, but have in common the use of, at least at some stages, the microfabrication technologies of the electronics industries. Though a brief description of the facilities required will be given, it is not within the scope of this chapter to deal with the details of the methods for standard microfabrication, or for micromask formation, which today usually involves computerized 'writing', e.g. electron-beam writing of electron-sensitive resists. It will be assumed that anyone embarking on adhesion micropatterning will have access to masks and to microfabrication facilities, and the protocols will reflect this. The electronics and physics departments of many universities have such facilities. Similarly, it will be assumed that experience of general cell culture and access to the appropriate facilities is available; only general background and non-standard cell culture methods will be detailed.

2. Basic photolithography

2.1 Overview

In simple micropatterning, the surface to be patterned is spin-coated with a thin layer of photosensitive polymer, a photoresist, which is then exposed to the appropriate illumination, and subsequently chemically developed to reveal the underlying substratum. In practice, the feature resolution of photolithography is 1–2 μm, though some laser–holographic methods allow submicron features to be defined. Electron-beam lithography, of the type used to write photomasks, can also be used to write submicron patterns directly on appropriate resists. This more specialized technique is probably less likely to be required for cell adhesion applications, and will not be considered further. Conventional microphotolithography will allow the micropatterning of any two-dimensional design. The basic facilities, which would normally be available in a photolithographic laboratory, would be as follows:

(a) The processes are carried out in a 'clean room' environment, i.e. a specialized laboratory where the temperature, humidity, and particle content of the air is tightly controlled. In many such facilities, particularly non-commercial research clean rooms, individual clean air cabinets, together with the use of particle-free materials (including water) and lint-free clothing, is sufficient for most purposes. These cabinets usually have compressed-air guns which dispense jets of filtered (i.e. particle-free) air to allow samples to be blown dry, thereby preventing drying marks. This environment is required since a crucial step in determining the quality of any pattern is dependent on good physical contact between the mask and the sample during exposure. Any particles will prevent good contact being made.

(b) Pre-defined photomasks of the desired micropatterns, which can be reused many times, are an absolute requirement. Many university photolithography facilities have mask-making capabilities. Custom made masks may also be obtained from commercial electronics companies, though this can be an expensive option. The two main styles of mask are dark field (transparent pattern in a dark background) and light field (dark pattern in a transparent background). Within a photolithography facility, masks designed for other purposes may be available for use for little or no cost.

(c) Vacuum spin-coating units hold samples onto a spinning chuck, and the chuck rotates from zero to thousands of r.p.m., linearly and in a fraction of a second. The viscoelastic properties of the photoresist solutions used in photolithography are such that they give a uniform coating on a flat substratum, when spun in this way. The thickness of the coating for a given resist is dependent on the maximum spin speed.

(d) Contact printer, which is a source of collimated light, usually blue or UV for most photoresists. The system is configured to allow exposure of a photoresist-coated surface through a mask in direct contact with it. This contact may be maximized by applying pressure, e.g. if the light source is below the sample, the best contact may be made by placing a weight on the sample. More sophisticated equipment, such as mask aligners, are designed to apply sufficient pressure for good contact to be made. Some facilities may use a projection system to transfer the mask pattern to the resist. In this instance, contact between the mask and sample does not take place, but clean and particle-free surfaces are still required so that the most efficient transfer can take place.

2.2 Primary pattern definition

Patterning photoresist on the desired substratum is the basis for most of the patterning techniques which will be described. The exact methods used will likely vary in different facilities. The most commonly used substrata for patterning adhesion are glass and silicon wafers. For patterning cell adhesion, silicon is less convenient as it is not transparent and therefore not compatible with standard microscopy for monitoring cell cultures. For this reason, the use of glass is preferable. For some applications, researchers may wish to have a surface whose chemistry is uniform and well defined; in this case fused quartz glass should be used. This is expensive, so if such defined chemistry is not a requirement, standard glass (e.g. a microscope slide) can be used. Here, a simple, flexible procedure which could be utilized in most microfabrication facilities, will be described. This will involve thorough cleaning of the substratum. Clean surfaces are important to maximize the adhesion of resist to the surface, and also to rid the surface of any particles. There are a number of methods for cleaning glass and silica surfaces. The protocol suggested is one which has been used successfully, though other methods are likely to be equally useful.

Protocol 1.   Cleaning substrata

Reagents

- Laboratory detergent
- Particle-free water

- Concentrated sulfuric acid
- 30% Hydrogen peroxide

Method

1. Cut the glass or silicon wafer to a convenient size (e.g. a half or a third of a standard microscope slide, though any other sizes compatible with the contact printer can be used). Safety note: always wear gloves and face protection when cutting glass or silicon.

2. Sonicate in a moderately concentrated solution of standard laboratory detergent (e.g. 0.5–2% (w/v) glassware cleaning detergent) for 10–15 min. (Solvents, such as methanol or acetone can be used to clean the samples instead of, or in addition to, detergent.)

3. Rinse thoroughly in clean room grade water, i.e. RO water which is particle-free. Drain of water.

4. In a glass or Teflon beaker, gradually add 30% hydrogen peroxide solution to concentrated sulfuric acid (to a maximum of one part peroxide to seven parts acid), stirring with a glass or Teflon rod, and monitoring the temperature with a glass thermometer. As the H_2O_2 is

---

Protocol 1. *Continued*

added, the temperature will rise. This should not be allowed to rise higher that 75–80 °C.[a]

5. Add the pieces of glass or silica to the acid/peroxide mixture and leave for 15–20 min. If a large batch of substrate pieces is being cleaned, then it is wise to gently disturb the pieces with the rod to ensure all surfaces to be cleaned are exposed to the mixture.

6. Tip the beaker into a larger container of clean water. Rinse the individual pieces in clean water and blow dry with filtered compressed air to remove all water droplets.

[a] This mixture is highly dangerous. The entire procedure must be carried out in a fume-hood, and appropriate protection should be worn.

The photoresist patterning process is summarized in *Figure 1A–E*. An important variable in this process is the resist used. Resists can be classed as positive (i.e. the areas illuminated are removed by development) or negative (i.e. the areas illuminated are retained on development), the former being more commonly used. The thickness of the spun resist layer depends both on resist type and on the maximum spin speed during coating. Within a broad, convenient range (1–2 μm), this thickness will have little effect on the procedures, though exposure times need to be increased with increasing thickness. One important parameter affected by resist thickness can be feature resolution. It should be noted that thinner (< 1 μm) resist layers may be required to resolve the smallest features. The following protocol is a standard procedure which will result in a uniform resist layer of approx. 2 μm. Thinner layers can be obtained using other resists, or by thinning the resist with the manufacturer's thinning agent. Since photoresists are sensitive to blue/UV wavelengths, the laboratory areas where this work is done are usually only illuminated with orange light, e.g. windows covered with orange filters, dark-room lights.

Protocol 2. Photoresist patterning (*Figure 1A–E*)

Reagents
- Shipley 1818 photoresist
- Particle-free water
- Shipley Microposit developer

Method

1. Place the cleaned, dried substratum (see *Protocol 1*) on the spin coater vacuum chuck.

2. Place drops of Shipley 1818 positive photoresist on the centre of the sample, avoiding bubbles in the solution. It is not necessary to completely cover the sample since the fluid will spread during spinning.

3. Spin at 4000 r.p.m. for 20 sec.

4. Remove sample from chuck and allow to stand for approx. 2 min or until a batch of samples has been coated, allowing the last sample to stand for 2 min.

5. Harden the resist layer by baking at 95°C for 30 min. Allow to cool before subsequent use.

6. Clean the photomask by sonication in acetone for 10 min, followed by rinses in clean water. Blow dry with filtered air.

7. Using the contact-printer, expose the photoresist through the mask. Exposure time will depend on the contact-printer type and wavelength used.

8. Develop in Shipley Microposit developer, diluted 1:1 with clean water, for 75 sec with gentle agitation.

9. Rinse in clean water. Blow dry with filtered air.

10. Examine the patterns under a light microscope. If exposed pattern is incomplete (i.e. development did not clear the exposed areas), then exposure time is too short. If resist is cleared from areas not intended to be exposed, or feature sizes in the resist are significantly larger than on the mask, then the exposure time may be too long.

3. Patterning organosilanes

Many of the methods for patterning adhesiveness have exploited the ability to covalently link organic molecules to glass and silica using organosilanes (2, 3, 10–16; see also Chapter 1). The two main classes of organosilanes used are alkylchlorosilanes and aminoalkylmethoxysilanes. Alkylchlorosilanes are highly reactive with glass and silica, and will form an immobilized film of alkyl chains on the reacted surface, this film being highly hydrophobic. Aminoalkyl-methoxysilanes will, under the right conditions, result in an immobilized film of aminoalkyl chains on the reacted surface, this surface being positively charged in an aqueous environment at neutral pH, and relatively hydrophilic. The protocols in this section will outline the method for patterning alkyl-chlorosilanes to produce a hydrophobic patterned surface on glass (*Figure 1E–G*). This pattern can then be subsequently used as the template for patterning aminoalkylmethoxysilanes, i.e. the aminoalkylmethoxysilanes will react with the unsilanated, non-hydrophobic areas (*Figure 2A,B*). These organosilanes can be volatile and noxious, and should be handled carefully in a fume-hood.

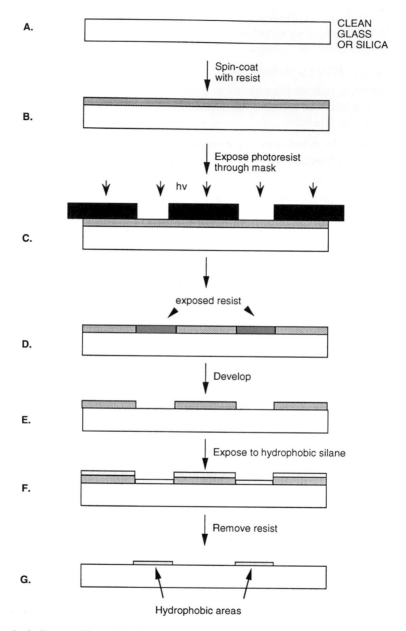

Figure 1. A diagram illustrating the steps of *Protocols 2* and *3* to produce micropatterned photoresist (A–E), and to use these to pattern alkylsilane (E–G). (hv) and arrows in (C) indicate exposure to light.

Protocol 3. Patterning alkylchlorosilane (*Figure 1E–G*)

Reagents
- Dimethyldichlorosilane
- Acetone
- Chlorobenzene

Method

1. Photoresist patterns on substrata fabricated using *Protocols 1* and *2* can be exposed to the alkylchlorosilane in two ways:

   (a) Immerse photoresist patterns in a 2% (v/v) solution of the alkyl-chlorosilane, dimethyldichlorosilane[a] (Sigma), in dry chloroben-zene, for 5 min. Rinse twice in chlorobenzene. Drain and carefully blow dry to remove remaining chlorobenzene.

   (b) Place photoresist patterns in a box (e.g. plastic sandwich box) along with a small watchglass containing 2–3 ml of the alkyl-chlorosilane, dimethyldichlorosilane. Replace the box lid and leave for 5 min (prolonged exposure to the organosilane vapour affects the photoresist and will lead to loss of pattern). Remove lid and patterns. Allow to stand in fume-hood to remove any residual surface vapour.

2. Remove the remaining photoresist by rinsing in three changes of acetone.

3. Rinse thoroughly in clean water. Blow dry.

[a] Other organosilanes, which will give a hydrophobic film, could be used, e.g. tridecafluoro-1,1,2,2,-tetrahydrooctyl-1-dimethylchlorosilane (3, 10).

Protocol 4. Patterning aminoalkylorganosilane (*Figure 2A,B*)

Reagents
- 2-Aminoethyl-3-aminopropyltrimethoxysilane
- Ethanol
- Acetic acid

Method

1. Immerse alkylorganosilane patterns, made using *Protocol 3*, in 1% (v/v) solution of 2-aminoethyl-3-aminopropyltrimethoxysilane[a] in 95% ethanol:5% water containing 1 mM acetic acid, for 10–15 min.

Protocol 4. *Continued*

2. Rinse thoroughly in anhydrous ethanol, followed by clean water. Blow dry.

3. Bake on a hot plate at 115–120°C for 10 min.

ᵃ Listed in the Aldrich chemical catalogue as 3-trimethoxysilylpropylethylenediamine. Other similar aminoalkylmethoxysilanes can be used, e.g. with one or three amino groups.

Another approach which has been successfully used to pattern cell adhesion by patterning organosilanes avoids the use of photoresists, but requires a laser with deep UV emission (3, 10). Briefly, surfaces coupled with amino-

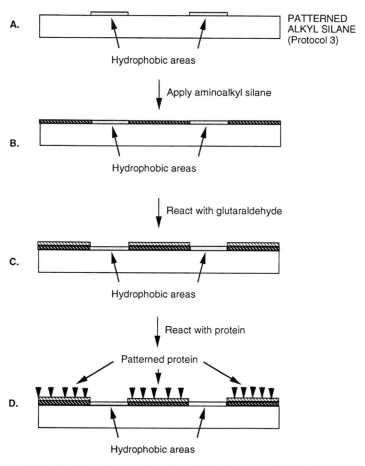

Figure 2. A diagram illustrating the steps of *Protocols 4* and *5* to produce micropatterns of alkyl- and aminoalkylsilanes (A and B), and the patterning of proteins by direct coupling to aminoalkyl silanated areas (B–D).

alkylorganosilanes (using *Protocol 4* on unpatterned surfaces) were exposed to deep UV from a laser source through a micropatterned mask. The exposed areas lose the coupled aminoalkyl chains and are therefore available for coupling hydrophobic chains using *Protocol 3*.

4. Cell culture on patterned substrata

4.1 Basic cell culture

Standard methods of culture of tissue cells involve the provision of an appropriate substratum. Tissue cells have an absolute requirement for such a substratum; they will not grow in suspension (some cancer cells will). A typical substratum is glass or tissue culture plastic (TCP) (TCP is modified polystyrene). Cells are usually grown in Petri dishes or flat-bottomed tissue culture flasks in a liquid nutrient medium. A small number of cells in suspension are added to the medium in such a vessel, and they will adhere, grow, and proliferate until they cover the available culture surface (i.e. confluence). The basal media used in cell culture contain defined mixtures of salts, amino acids, vitamins, and sugars. They are usually supplemented with blood serum, particularly bovine serum, though sera from other sources are used (e.g. horse, human). This serum supplement provides essential agents absent in the defined component of the medium. These agents include hormones, growth factors, and attachment factors. Serum-derived attachment factors will passively adsorb to culture surfaces. Cells in standard culture are actually adhering to adsorbed serum-derived attachment factors, the main attachment factors in sera being the glycoproteins, vitronectin, and fibronectin. Alternatively, cells can be cultured in serum-free conditions if the basal medium is supplemented with insulin, the iron transporting molecule, transferrin, and selenium ions. In serum-free culture, the culture surface is usually pre-coated with attachment factors. These may be purified serum-derived attachment factors, or other extracellular matrix (ECM)-derived molecules such as collagens, gelatin, or laminins. Cells are usually spherical in suspension (suspensions are prepared by using proteolytic enzymes, such as trypsin, to dissociate tissues or cultures), but they normally spread, or extend elongating processes (e.g. axons of neurones), on appropriate attachment factors. Although cells can adhere to 'non-physiological' surfaces (e.g. in the absence of attachment factors cells will adhere to positively charged surfaces, such as poly-L-lysine coated glass or plastic), they usually will not spread, be metabolically active, or proliferate. For these reasons, any attempt to pattern adhesiveness for cells will ultimately require the patterning of attachment factors.

For a more detailed consideration of cell culture, consult one of the many fine methods books which are available, e.g. *Animal cell culture: a practical approach*, 2nd edn (ed. R. I. Freshney).

The selective coating of attachment factors to patterned surfaces may occur by two means:

(a) Attachment factors may preferentially adsorb to chemically patterned surfaces, and therefore cells will preferentially adhere to and locomote on these areas (*Figure 3*).

(b) A surface may be patterned with chemical groups to which proteins, such as attachment factors, can be covalently cross-linked. This will lead to a pattern of immobilized protein to which cells preferentially adhere.

The following protocols will use both preferential adsorption and patterned immobilization of attachment factors. Since the adhesion of cells to different attachment factors can be dependent on cell type, care must be taken when choosing the attachment factor(s) to be patterned. For example, some neurones derived from the central nervous system will only poorly adhere to serum-derived attachment factors, and will not extend axonal and dendritic processes unless the surface has been pre-coated with the ECM protein, laminin.

4.2 Patterning adhesion by adsorption of attachment factors

The simplest approach (*Protocol 5*) which may be sufficient for many cell types will be to culture the cells in serum-containing medium, on patterned alkylorganosilanes made using *Protocol 3*. These will essentially be patterns of hydrophobicity on untreated glass which have been shown to be differentially adhesive for fibroblastic and epithelial cells (11, 12). The serum-derived attachment factors, vitronectin and fibronectin, adsorb well to untreated glass, but not to hydrophobic surfaces. Therefore, cells will adhere to the untreated areas of the pattern. It has been shown that these attachment factors will slowly adsorb to the hydrophobic areas, but the adhesiveness of these areas remains lower than on untreated glass (11). Cell culture reagents, including media, sera, and purified attachment factors, are available from many well known suppliers, including Gibco BRL, Flow Laboratories, and Sigma.

Protocol 5. Cell culture of patterned hydrophobic silanes (*Figure 3A,B*)

Reagents
• Serum-containing cell culture medium

Method
1. Use standard enzymatic dissociation techniques to prepare a cell suspension. Determine the concentration of the cells using a cell counting chamber.

---

2. Place the hydrophobic patterned surface (*Protocol 3*)[a] face up in a Petri dish, and add culture medium, which contains serum,[b] containing cells at moderate concentration.[c]

3. Incubate at the appropriate temperature (37 °C for most mammalian and avian cells) for 2–3 h.

4. Remove unattached cells by gently agitating the dish and removing the medium.

5. Replace the medium with fresh, pre-warmed, cell-free medium.

6. Examine cultures with inverted phase-contrast microscopy. Patterned attachment may be evident. Incubate further, at least another 24 h, to allow complete spreading and the establishment of the culture.

---

[a] Patterns produced using *Protocol 4* may also be used for this protocol. The relatively hydrophilic and charged surface offered by the aminoalkyl silanated surface appears to readily adsorb attachment factors.
[b] A typical culture medium consists of the basal medium supplemented with 10% (v/v) fetal bovine serum or new-born calf serum.
[c] The number of cells required will depend on the area of the culture surface, which will be the total dish area, if the cell suspension is initially poured over the whole dish. A moderate concentration would be 20 000 cells/cm². For some purposes, higher or lower densities may be required.

It would be expected that, because many proteins, including serum-derived attachment factors, preferentially adsorb to untreated glass, as opposed to the hydrophobic silane surfaces, other ECM-derived attachment factors might behave similarly. This is not the case for purified laminin. When solutions of laminin were applied to patterned hydrophobic silanes (*Protocol 3*), it was found to have preferentially adsorbed to the previously hydrophobic regions (13). These areas have been found to preferentially support the adhesion, growth, and survival of brain neurones and muscle cells (*Figure 3C,D*) (13, 15).

Protocol 6.   Patterning laminin on hydrophobic silanes
                 (*Figure 3C,D*)

Reagents
- EHS laminin
- PBS

Method

1. Prepare a 5 μg/ml solution of EHS laminin (Sigma) in phosphate-buffered saline (PBS) pH 7.4.

2. Apply the laminin solution to the surface of a hydrophobic pattern (*Protocol 3*).[a]

Peter Clark

Protocol 6. *Continued*

3. Incubate at 37 °C for 2–3 h.

4. Remove laminin solution and rinse in PBS.

5. Seed cells onto the pattern as in *Protocol 5*.

ᵃ Because areas of the patterned surface are highly hydrophobic, care must be taken to ensure that the aqueous laminin solution covers the entire sample. Samples may be completely immersed in excess laminin solution in a Petri dish (this can use large amounts of expensive laminin), or enough solution can be applied only to the patterned surface, making sure that the hydrophobic regions do not repel it.

4.3 Patterning adhesion by covalent immobilization of proteins

Proteins can be covalently cross-linked to aminoalkylsilane derivatized glass surfaces using glutaraldehyde as the linking molecule. The method was origin-

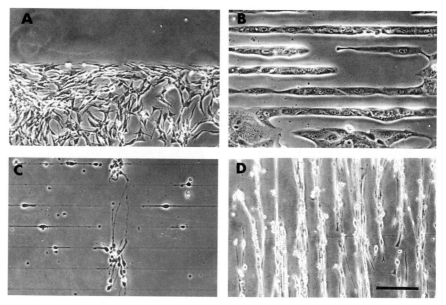

Figure 3. Phase-contrast micrographs of cells cultured on micropatterned glass surfaces. (A) BHK fibroblasts, cultured on alkylsilane patterned glass in the presence of serum, fail to adhere to, or migrate onto, the hydrophobic regions of the pattern (the upper area) (11). (B) MDCK epithelial cell, cultured on surfaces consisting of alternating tracks of alkyl silanated and unmodified glass, in the presence of serum, on which the cells form linear colonies (12). (C) Chick embryo brain neurones cultured, in serum-free medium, on alkylsilane patterns pre-coated with laminin; the laminin preferentially adsorbs to the previously hydrophobic alkylsilane areas (a wide vertical bar with narrow, 2 μm wide tracks perpendicular to it); the neurones adhere to, and extend neurites on, these areas (13). (D) Mouse muscle cells (myoblasts) preferentially adhere, differentiate into myotubes, and survive, on wide laminin tracks (as in C) (15). The bar in (D) represents 200 μm in (A–C), and 400 μm in (D).

106

ally devised for studying cell adhesion (16), and has been combined with silane patterning to provide patterned adhesiveness (8, 17). The basic method for protein immobilization, modified by Enrique Perez-Arnaud in Geoffrey Moores laboratory (17), is given in *Protocol 7* (see also *Figure 2B–D*). The original method (16) used sodium borohydride at the end of the procedure to reduce the Schiff's bases formed by the reactions of the aldehyde groups with the amino groups, thereby stabilizing the links. The modified method uses sodium cyanoborohydride at the two linking stages. This is a less harsh method which is more likely to spare proteins which are sensitive to such treatments.

Protocol 7. Immobilization of proteins on aminoalkylsilane patterns (*Figure 2B–D*)

Reagents
- Glutaraldehyde
- Sodium cyanoborohydride
- PBS
- Bovine serum albumin or caesin

Method

1. Immerse the pattern (produced using *Protocol 4*) in a solution of 2% glutaraldehyde in PBS pH 7 containing 40 mM sodium cyanoborohydride ($NaCNBH_3$) for 1 h, at room temperature.

2. Rinse twice in distilled water.

3. Add the protein solution[a] (20–200 μg/ml, depending on amount required on surface) in PBS with 40 mM $NaCNBH_3$, for 1 h at room temperature, or 16 h at 4 °C.

4. Rinse two or three times in PBS.

5. Incubate for 1 h in a 200 μg/ml solution of an irrelevant (i.e. non-adhesive) protein,[b] such as bovine serum albumin (Sigma) or casein (Sigma) in PBS with 40 mM $NaCNBH_3$.

6. Rinse in PBS. The surface is ready for cell culture.

---

[a] One important consideration is that when this protocol is used to pattern a protein on aminoalkylsilane/alkylsilane patterns, as produced by *Protocol 4*, passive adsorption of the protein to the hydrophobic surface may also occur. For example, laminin may not be appropriate to use is this way because it can preferentially adsorb to hydrophobic silane glass (13). If protein adsorption to the hydrophobic surface proves to be a problem, the differential adhesiveness of such patterns could be lost. This could be restored by washing the patterns in solutions (e.g. high salt or urea) which will 'elute' the non-covalently linked protein leaving a protein pattern. This removal procedure must not affect the integrity or function of the bound proteins.
[b] This step is particularly important when low concentrations of the original protein are used. The irrelevant protein will block unreacted aldehyde, preventing inappropriate binding during cell culture.

Derivatization of irrelevant, non-adhesive proteins may be used to pattern non-adhesiveness. Subsequently, an attachment factor may be preferentially adsorbed, or derivatized, to the areas devoid of protein, thereby providing differential adhesiveness. This approach was used by Perez-Arnaud (17) to pattern laminin with adjacent non-adhesive bovine serum albumin, which provided a differentially adhesive surface guiding neurite outgrowth of neurones. The method involves patterning photoresist on a surface which had been derivatized with aminoalkylsilane, coupling non-adhesive protein to the exposed silane, removing the photoresist to expose uncoupled silane, to which attachment factors can be adsorbed or derivatized.

Protocol 8. Patterning 'non-adhesiveness' using protein derivatization

Reagents
- 2-Aminoethy-3-aminopropyl trimethoxysilane
- Bovine serum albumin or casein
- Shipley 1818 photoresist
- Acetone

Method

1. Derivatize an unpatterned, clean glass surface with an aminoalkylsilane using *Protocol 4*.

2. Pattern photoresist on this surface, using *Protocol 2*, thereby exposing a pattern of aminoalkylsilane.

3. Couple a non-adhesive protein (e.g. bovine serum albumin or casein, both from Sigma) using *Protocol 7*.

4. Remove the patterned photoresist by rinsing in acetone, as in *Protocol 3*, step 2.[a]

5. Rinse in PBS. This surface should be differentially adhesive for culturing cells in the presence of serum or by pre-coating with laminin which should preferentially adsorb to the aminoalkyl surface.

[a] The acetone rinse may denature the immobilized protein. This is of no consequence since it is patterned in order to block adhesion.

---

5. Photopatterning cell-attachment molecules

A method has recently been described which allows the sequential patterning of proteins (18, 19). This technique exploits the strong, specific interaction between the protein, streptavidin (or a modified avidin), and biotin (see Chapter 2). In this system, an analogue of biotin which has been functionalized with a photoactive group, photobiotin, is bound to an avidin-modified surface. Protein is patterned onto this surface by exposing selected areas of a film of protein solution on the modified surface, to light through a mask, activating the photosensitive group which will form a bond with the protein. This procedure can be used to pattern attachment factors. Indeed with appropriate masks and mask aligning capability, it is possible to fabricate a pattern of a number of spatially separated proteins (19).

Protocol 9. Patterning adhesion using photolabile surface derivatization

Reagents
- Neutravioin
- Photobiotin
- PBS
- Bovine serum albumin or casein

Method

1. Derivatize an unpatterned, clean glass surface with an aminoalkyl-silane using *Protocol 4*.

2. Couple Neutravidin (Pierce and Warriner) to the aminoalkyl surface from a 200 µg/ml solution using *Protocol 7*, including the blocking procedure (steps 5 and 6).

3. Incubate the sample in a 10 µg/ml long arm photobiotin[a] (Vector Laboratories) in PBS for 20 min. Rinse in PBS.

4. Cover the sample with the protein[b] solution, 10 µg/ml in PBS, and expose to light[c] through a mask. Rinse in PBS.

5. To block unreacted photobiotin sites, apply a solution of irrelevant protein (e.g. bovine serum albumin or casein) at 2 mg/ml.

6. Flood expose to light (i.e. no mask). Rinse in PBS. The sample may be used as a surface for cell culture.

[a] This and all subsequent steps must be carried out in the dark, unless exposure to light is specified.
[b] The choice of attachment factor will depend on the cell type(s) used and may also be limited by the sensitivity of certain proteins (possibly laminin) the wavelengths used to photopattern.[c]
[c] A mercury vapour lamp is suitable for this purpose. Photobiotin is activated by 340–375 nm wavelengths. A filter which removes wavelengths below 300 nm (e.g. Hoya SL glass filter) may be desirable to prevent UV inactivation of the various components.

---

6. Other approaches

Space does not allow detailed consideration of other methods which have been successfully applied to patterning cell adhesiveness, but these methods may be more appropriate to researchers in certain fields, and will therefore be briefly described.

Patterned metal oxides (e.g. SnO_x, InO_x) provide a preferentially adhesive surface for the growth of neuronal cell processes, this having been correlated with the electronegativity of the oxides (6). These patterns can be made a using a 'lift-off' method: the oxides are deposited on photoresist pattern (*Protocol 2*), and when the photoresist is removed, a pattern of metal oxide remains.

Microphotolithography has been used to fabricate masks through which metals can be evaporation deposited. Deposition of palladium metal, through such a mask, onto culture surfaces coated with the polymer, polyhydroxyethylmethacrylate (poly HEMA), will produce adhesive areas (the metal) on a non-adhesive background (poly HEMA) (1, 20). This type of pattern has been shown to be differentially adhesive to fibroblasts cultured in serum-containing medium.

An ingenious method of patterning gold coated surfaces uses an elastomeric stamp to pattern non-adhesiveness (21). The stamp is made by applying silicone elastomer (Dow Corning) onto a photoresist template (*Protocol 2*), the cured silicone being subsequently peeled away to produce a surface having the mirror relief of the photoresist pattern. This stamp is then 'inked' with an alkylthiol and applied to the gold surface. The alkyl groups will be bound to the gold surface in the pattern of the stamp, this pattern acting as a template for hydroxyalkylthiol patterning, to which extracellular matrix proteins, such as laminin, will preferentially adsorb. Chapter 4 describes the interaction of thiols with gold surfaces in more detail.

A quick and simple method of patterning laminin to a degree which may be useful for some studies, and which does not require specialist equipment or micromasks, was described by Hammarback *et al.* (22). This method employs UV inactivation of laminin adhesiveness. Laminin is passively adsorbed to a culture surface (pre-coating with 10 µg/ml poly-L-lysine (Sigma) in PBS enhances laminin adsorption). A mask is placed to cover areas destined to be adhesive. The mask may be anything which gives a convenient pattern, e.g. electron microscope grids of various mesh sizes have provided useful patterns (22). The surface is exposed to deep UV irradiation (a mercury lamp can provide appropriate wavelengths). The irradiated areas lose laminin-mediated adhesiveness, such that when neurones are cultured on these patterns, neurite outgrowth is confined to the unirradiated areas.

Acknowledgements

I wish to thank Bill Monaghan, Mary Robertson, Joan Carson, Lois Hobbs, Chris Wilkinson, and other staff at the Department of Electronics and Electrical Engineering of the University of Glasgow, for their generosity with their time and resources. I also thank Geoffrey Moores, Adam Curtis, Steve Britland, Hywel Morgan, Philip Hockberger, and Don Ingber for providing reprints, pre-prints, and other information.

References

1. O'Neill, C., Jordan, P., and Ireland, G. (1986). *Cell*, **44**, 489.
2. Kleinfeld, D., Kahler, K. H., and Hockberger, P. E. (1988). *J. Neurosci.*, **8**, 4098.
3. Dulcey, C. S., Georger, J. H., Krauthamer, V., Stenger, D. A., Fare, T. L., and Calvert, J. M. (1991). *Science*, **252**, 551.
4. Clark, P. (1996). In *Nanofabrication and biosystems* (ed. H. C. Hoch, L. W. Jelinski, and H. G. Craighead), p. 356. Cambridge University Press, New York.
5. Hockberger, P. E., Lom, B., Soekarno, A., Thomas, C. H., and Healy, K. E. (1996). In *Nanofabrication and biosystems* (ed. H. C. Hoch, L. W. Jelinski, and H. G. Craighead), p. 276. Cambridge University Press, New York.
6. Kawana, A. (1996). In *Nanofabrication and biosystems* (ed. H. C. Hoch, L. W. Jelinski, and H. G. Craighead), p. 258. Cambridge University Press, New York.
7. Clark, P., Connolly, P., Curtis, A. S. G., Dow, J. A. T., and Wilkinson, C. D. W. (1991). *Sensors and Actuators*, **B3**, 23.
8. Britland, S. T., Perez-Arnaud, E., Mc Ginn, B., Clark, P., Connolly, P., and Moores, G. R. (1992). *Biotechnol. Prog.*, **8**, 155.
9. Clark, P. (1994). *Biosens. Bioelectron.*, **9**, 657.
10. Bhatia, S. K., Teixeira, J. L., Anderson, M., Shriver-Lake, L. C., Calvert, J. M., Georger, J. H., *et al.* (1993). *Anal. Biochem.*, **208**, 197.
11. Britland, S. T., Clark, P., Connolly, P., and Moores, G. R. (1992). *Exp. Cell Res.*, **198**, 124.
12. Clark, P., Connolly, P., and Moores, G. R. (1992). *J. Cell Sci.*, **103**, 287.
13. Clark, P., Britland, S., and Connolly, P. (1993). *J. Cell Sci.*, **105**, 203.
14. Lom, B., Healy, K. E., and Hockberger, P. E. (1993). *J. Neurosci. Methods*, **50**, 385.
15. Clark, P., Coles, D., and Peckham, M. (1997). *Exp. Cell Res.*, **230**, 275.
16. Aplin, J. D. and Hughes, R. C. (1981). *Anal. Biochem.*, **113**, 144.
17. Perez-Arnaud, E. (1994). *A method for patterning proteins and its application to study the guidance of neurite outgrowth*. Ph. D. Thesis, University of Glasgow.
18. Pritchard, D. J., Morgan, H., and Cooper, J. M. (1995). *Angew. Chem. Int. Ed. Engl.*, **34**, 91.
19. Pritchard, D. J., Morgan, H., and Cooper, J. M. (1995). *Anal. Chem.*, **67**, 3605.
20. O'Neill, C., Jordan, P., Riddle, P., and Ireland, G. (1990). *J. Cell Sci.*, **95**, 577.
21. Singhvi, R., Kumar, A., Lopez, G. P., Stephanopoulos, G. N., Wang, D. I. C., Whitesides, G. M., *et al.* (1994). *Science*, **264**, 696.
22. Hammarback, J. A., Palm, S. L., Furcht, L. T., and Letourneau, P. C. (1985). *J. Neurosci. Res.*, **13**, 213.

References

1.
2.
3.
4.
5.
6.
7.
8.
9.

7

Sol-gel matrices for protein entrapment

BAKUL C. DAVE, BRUCE DUNN, JOAN S. VALENTINE, and
JEFFREY I. ZINK

1. Introduction

Traditional inorganic glasses are not generally considered to be a suitable host matrix for immobilization of biological molecules. Consequently, stabilization of proteins and enzymes within inorganic silica glasses is very unusual (1). This breakthrough in immobilization methods has been made possible by the sol-gel method of synthesizing inorganic solids (2). The sol-gel approach is a solution-based synthesis technique for making inorganic materials at low temperature. The main advantage of the sol-gel method is that inorganic oxide glasses are synthesized by hydrolysis of simple alkoxo precursors (3). In the case of silica-based systems, the hydrolysed precursors then condense together to form a predominantly SiO_2 glass framework. In this way, SiO_2-based silica glass can be conveniently fabricated using $Si(OR)_4$ precursors at room temperature without the extreme temperatures required for conventional glass melting methods. Sol-gel-derived silica glasses are highly porous structures containing an interstitial solvent phase. A high molecular weight dopant molecule added to the initial reaction mixture becomes trapped in the interconnected porous network as the final SiO_2 glass structure forms (4). In addition, since the pore diameters are much less than the wavelength of light, the sol-gel glasses are optically transparent. This enables one to monitor spectroscopically the nature of events associated with encapsulation and reactivity of the dopant molecules (5).

The sol-gel-derived silica glasses offer unique advantages for stability and reactivity of biomolecules (6). An important feature of the sol-gel materials which retain an aqueous phase in the porous structure is that it allows one to incorporate and immobilize biological molecules in the glass. The immobilization is due to physical encapsulation within the porous structure without any covalent interaction, and a functional modification of the protein is not required. A variety of different proteins can, thus, be immobilized in the sol-gels (1). Physical encapsulation leads to isolation of individual molecules and prevents self-aggregation effects. The smaller pores of the matrix render the immobilized protein insusceptible to microbial degradation since the bacteria are unable to penetrate the sol-gels. The cage effect provided by physical encapsulation within a pore also prevents unfolding and denaturation of the encapsulated protein. At the same time, the porous matrix allows the immobilized biomolecule to react with low molecular weight substrates which are able to diffuse in and out of the material and are also freely mobile within the matrix (4).

This chapter is concerned primarily with elaborating different practical aspects of encapsulation, stability, and reactivity of encapsulated biological molecules in sol-gel-derived porous host media. A variety of proteins, enzymes, and other biosystems are functionally active within these glasses (1). In almost all the cases investigated so far, the chemistry of dopant biomolecules in gel glasses is analogous to that in solution with the exception that now the system involves a porous silica sol-gel matrix. In this way, a heterogeneous, multiphase reaction chemistry of proteins and enzymes is designed to occur in the porous structure of the inorganic host sol-gel matrix. The unique state of aggregation of such sol-gel composites, where dopant biomolecules are dispersed in the nanopores of the material, exemplifies a state intermediate between the isotropic solution phase and the solid state. These systems utilize the properties of matrix encapsulated biomolecules in a solvent-rich environment and different proteins and enzymes can be made to carry out various processes such as ligand binding and catalysis to generate specific chemical responses. Various functional biocatalysts can be prepared using the sol-gel method of immobilization. This chapter describes and reviews techniques for encapsulation of different proteins and enzymes in sol-gel-derived silica materials. Specifically, the effects of encapsulation and reaction chemistry of various dopant biomolecules in these matrices are discussed.

2. Overview of the sol-gel process

The sol-gel process is a low temperature method of making glass (3). The manufacture of glass by high temperature melting of silica has been known for centuries. However, due to the extremes of the temperatures required, the glass matrix prepared by this route is not feasible as a host matrix to a majority of organic and inorganic dopants. Recent advances in the molecular chemistry of silica have made possible alternative synthetic pathways for making glass that are feasible at room temperatures. The advent of the sol-gel process for making silica glass has made the matrix suitable for microencapsulation of a variety of molecules, most notably biomolecules, which are extremely sensitive to thermal denaturation. The formation of the silicate matrix is achieved by hydrolysis of an alkoxide, usually tetramethyl orthosilicate (TMOS), followed by condensation to yield a polymeric oxo-bridged SiO_2 network. Hydrolysis results in conversion of Si–OR bonds to Si–OH bonds which condense to form an oxo-bridged polymeric Si–O–Si structure. These reactions occur in a localized region and lead to the formation of sol particles. As polycondensation continues, the viscosity of the sol begins to increase. This viscous material then solidifies and leads to formation of a porous gel. The resulting sol-gel material is an aqueous solid containing up to 70% water by weight.

2.1 Reaction chemistry

In the sol-gel process, glass formation occurs via hydrolytic polymerization of mononuclear precursors. The preparation of inorganic glasses by the sol-gel route involves a molecular precursor which can be hydrolysed, leading to the formation of an inorganic oxide network. Although such a network is formed by a variety of elements, e.g. Si, V, Ti, Al, Zr, etc. (7), silicates have been the most thoroughly studied and characterized systems. The formation of a silicate matrix is usually achieved by the hydrolysis of an alkoxy precursor to yield a polymeric oxo-bridged SiO_2 network. The chemical reactions taking place are summarized in *Figure 1*. As can be seen from the reaction, the process liberates molecules of alcohol corresponding to the alkoxide used. For this reason, the methoxy precursor TMOS is used for the encapsulation of biomolecules as they tend to denature in higher homologues of alcohols.

The nature of the matrix can be controlled synthetically. The formation of glass by the sol-gel method is a sequential process of polymerization from a simple monomeric precursor, and to some extent there is control over the product stoichiometry and geometry. The relative rates of hydrolysis and condensation govern gel morphology (8). As can be seen from the two-step nature of the reaction, the composition of the final product should depend

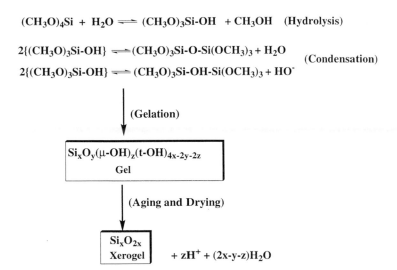

Figure 1. Chemical reactions occurring during sol-gel–xerogel structural transformation.

upon the rates of the individual reactions. A higher rate of hydrolysis would lead to replacement of alkoxy ligands of the precursor complex with hydroxy groups, resulting in extensive condensation, and yielding a highly branched interconnected porous network.

Another variant which determines the stoichiometry of the final gel is the pH of the reaction medium (8). Acid catalysis leads to materials with smaller pores whereas base catalysis leads to structures with larger pores (9). While pH does not directly influence the reaction, a catalytic agency of protons or hydroxide ions is required for individual reaction steps to propagate at acceptable rates. If the reaction medium is acidic, then the influence of the proton on the reaction is twofold: a faster rate of hydrolysis and a slower rate of polycondensation. The presence of free protons in the medium affords protonation of the ligated alkoxide and thereby favours its dissociation. On the other hand, an acidic medium hinders the formation of oxo-bridges resulting from the loss of protons from water or hydroxide molecules. Therefore, the overall influence of acidic catalysis is to increase the rate of hydrolysis and to disfavour condensation reaction. Base hydrolysis causes the opposite effect. The presence of hydroxide anions in the reaction medium facilitates the formation of oxo-bridges and therefore speeds up the rate of condensation.

Hydrolysis and condensation of the TMOS precursor leads to formation of smaller aggregates of silica particles that remain suspended in the reaction medium. The solution containing these smaller particles is called a sol. Stability of the sol depends on the relative rates of hydrolysis and condensation. The sol particles combine together to yield an extended three-dimensional network. When the degree of cross-linking increases, the sol turns more viscous and finally turns to a solid gel. The point at which the liquid to solid transition takes place is termed the gelation point.

Even after the gelation point, the structure and properties of the gel continue to change as long as its pore liquid is not allowed to dry (10). In this 'aging' process there is continuing polycondensation reactions in the solid amorphous phase that increase cross-linking of the silicate network. Another process that takes place is the spontaneous shrinkage of the gel and resulting expulsion of pore liquid from the gel. This expulsion of liquid is primarily due to the formation of new Si–O–Si bonds via polycondensation of Si–OH fragments that result in contraction of the gel network. Due to these processes taking place during aging, the strength of the gel increases and pore sizes in the gel become smaller. Most of the solvent water and the methanol generated during hydrolysis/condensation remains in the gel producing an aged gel which is a solid state glassy material with a trapped solvent phase. An aged gel is thus a two phase system comprised of a porous inorganic solid and a trapped aqueous phase. The sol-gel materials exhibit the following characteristics:

- polymeric silicate network structure
- mechanical rigidity

- optical transparency above 300 nm
- pore size distribution with an average pore diameter $\sim$ 100 nm
- retention of solvent phase within the interconnected porous structure

When the pore liquid of the gel is allowed to evaporate, the gel volume decreases due to loss of liquid, and the gel shrinks. As the evaporation of the solvent takes place, capillary forces induce the gel network to draw together producing smaller pores. The ambiently dried gels, termed xerogels, are approximately 70% their original size, show no further loss of pore liquid, and are structurally invariant. During the aged gel to xerogel structural transformation the following changes occur:

- expulsion of water
- loss of –OH groups to form Si–O–Si linkages
- pore collapse
- mechanical strengthening
- bulk volume shrinkage by $\sim$ 70%

The xerogels are mechanically much more robust than aged gels due to greater density and formation of an extensive silicate network. A xerogel is a dimensionally stable glass.

Protocol 1. Synthesis of TMOS-derived SiO_2 sol

Equipment and reagents
- Ultrasonic bath
- Tetramethyl orthosilicate (TMOS) (Aldrich)
- 0.04 M hydrochloric acid

Caution: carry out all operations in a well ventilated fume-hood.

Method

1. Weigh out 15.27 g of TMOS in a 100 ml beaker.

2. Add 3.39 ml of deionized water.

3. Add 0.22 ml of 0.04 M HCl solution.

4. Transfer the beaker to an ultrasonic bath containing ice.[a]

5. Sonicate the mixture in the bath for 15–20 min until the mixture turns homogeneous to yield sol.

6. Store the sol in an ice-bath or in a refrigerator at 4°C.

[a] Caution: fire hazard. The flash point of TMOS is 28°C and ice is necessary to ensure safety.

2.2 Processing

The flexible solution pathway for making sol-gel materials enables one to process the resulting materials in different configurations. Two processing methods based on bulk and thin films are described.

2.2.1 Bulk processing

The processing of bulk gels (or monoliths) offers unique advantages to obtain optical quality solids. Monoliths are sol-gels with dimension ≥ 1 mm that are cast to a desired shape and processed without fracture. A variety of different shapes can be obtained by transferring the sol to appropriate containers and allowing gelation. As the sol turns to a solid gel it assumes the shape of the container and in this way many complex shapes can be obtained. For the sol-gels containing immobilized biomolecules, processing of the gels is often carried out in 4 ml cuvettes. The gels formed in a cuvette are suitable for optical absorption spectroscopy measurements. The processing steps for monolithic sol-gels are shown in *Figure 2*. In order to obtain bulk monoliths, the sol containing the dopant protein or enzyme is transferred to a polystyrene cuvette. After gelation, a solid material is formed in the cuvette. The freshly formed sol-gels are covered with Parafilm and are allowed to age under sealed conditions for two to three days. The aging may be carried out at room temperature or, alternatively, at 4°C in a refrigerator depending on the dopant molecule. During this aging period the gels shrink and can be removed from the cuvettes.

The aged gels are chemically and physically dynamic solids and careful handling of the samples is required to prevent further evolution of the material caused by loss of solvent and drying. Drying induces collapse of the porous structure within the material and gives rise to capillary stresses. Carefully controlled solvent evaporation from the aged gel is critical as the structural integrity of the gels may degrade due to excessive internal stresses. The aged gels left in air usually crack irreversibly within minutes. In order to prevent fracture, the aged gels must be kept wet at all times. A quick transfer of the gels to another medium with minimal solvent loss is vital to preserve integrity of the samples. In this way, the aged gels may be washed with suitable buffer solutions to maintain pH and to remove excess methanol. The sol-gels that are aged for only a few days are mechanically very fragile and must

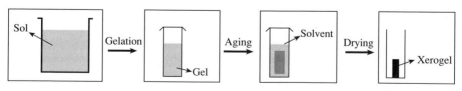

Figure 2. Schematic of physical changes associated with processing of sol-gel monoliths.

be handled carefully to avoid fracture. Longer aging periods (three to four weeks) usually result in sol-gels with improved mechanical properties.

The aged gels may also be dried under ambient conditions to obtain stable xerogels. Extensive stresses caused by non-uniform drying usually lead to catastrophic fracture of the materials. In order to avoid cracking, slower drying is crucial. The strategy employed for slower drying is based on controlled solvent loss from the sol-gels such that equilibrium conditions are maintained. A slow solvent evaporation (over a period of six to eight weeks) through small pin-holes made in the Parafilm covering usually results in excellent optical quality monoliths. The ambiently dried sol-gels are structurally stable materials.

2.2.2 Thin film processing

The processing of sol-gel-derived thin films is very appealing from a techno-logical viewpoint (11). Thin films can be formed on a suitable substrate from a low viscosity sol by dip coating, spin casting, or spraying (12). Sol-gel tech-niques for processing of optical quality thin films of silica glass are well estab-lished (13). Thin film-based materials are attractive due to several salient features including:

- better dimensional control of final product material
- smaller amounts of dopant biomolecules required
- faster response times with external reagents
- better compatibility with optics and electronics
- the ability to fabricate arrays or multilayer configurations

The dip coating method of thin film deposition is perhaps more important technologically since a uniform coating can be deposited onto substrates of large dimensions and complex geometries (14). In the dip coating process, a suitable substrate is withdrawn slowly at a constant speed (typically 10–20 cm/min) from a low viscosity sol (15). The arrangement used for dip coating of films is shown in *Figure 3*. In order to deposit uniform homogeneous films, a steady vibration-free withdrawal of the substrate from the sol is necessary. In the authors' laboratory this is accomplished by a simple, inexpensive mech-anism provided by hydraulic motion (*Figure 3*). Glass microscope slides or polished silicon wafers can be used as substrates. A Mylar ribbon is used as a guiding tool that is attached on one end to a float and on the other end to the substrate. The substrates are attached to one end of the pulley system by using binder clips (or adhesive tape) and withdrawn under a fluctuation-free hydraulic pressure provided by a float kept in a draining water reservoir. The rate of withdrawal is adjusted by controlling the drainage flow rate through a drain valve.

The deposition of the sol onto the substrate is due to non-covalent inter-actions such as H bonding and electrostatic attraction between the surface of

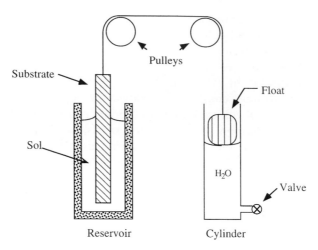

Figure 3. Schematic of apparatus used for dip coating of sol-gel thin films.

the substrate and the sol. The formation of a thin film from deposited sol is aided by gravitational draining, and the conversion of deposited sol to porous gel is caused mainly by solvent evaporation. Loss of solvent molecules accelerates hydrolysis and condensation steps and leads to the formation of a porous SiO_2 gel structure.

The thickness of these films is governed by the viscosity of the sol while their homogeneity depends on the rate of withdrawal. The entrained silica sol particles are concentrated due to solvent evaporation and compacted by capillary forces. The structural evolution in sol-gel thin films is very complex. Unlike the bulk gel system where gelation, aging, and drying occur sequentially over a period of several weeks or longer, all of these processes typically occur within 30 seconds in the thin film such that the drying stage overlaps the gelation and aging stages. The important microstructural characteristics of the dip coated sol-gel films such as porosity, surface area, and pore size are determined by pH and viscosity of the initial sol, the rate of film pulling, the ambient temperature, the rate of drying, the relative rates of evaporation and condensation during drying, and the magnitude of drying-induced capillary forces (16).

3. Sol-gel encapsulation of biomolecules

The sol-gel materials can be used as host matrices for a variety of biological molecules (1, 17). The biomolecules added to the sol get trapped within the growing polymeric network as the final porous gels form (*Figure 4*). The dopant biomolecules reside in the porous network of the sol-gel-derived matrix (6). The sol-gel immobilized biomolecules retain their reactivity upon encapsulation and are able to bind low molecular weight ligands as well as

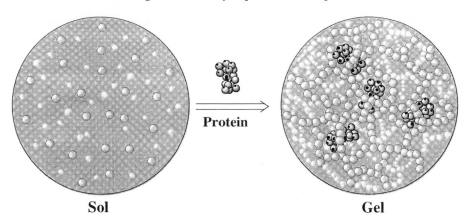

Figure 4. Immobilization of protein during formation of solid gel from sol.

carry out catalysis of substrates (17, 18). The molecular sizes of substrates and ligands are considerably smaller than the average pore diameters and thus they are mobile in the solvent-filled pores (4). The high molecular weight biomolecules, on the other hand, are confined within the pores of the sol-gel matrix.

The sol-gel procedures generally employed for encapsulation of organic and organometallic molecules employ pH values and methanol concentrations that are rather unsuitable for immobilization of proteins and enzymes which tend to denature under these conditions. To stabilize fragile biological molecules in these glasses, the strategy that has been quite successful is that of adding buffer to the initial sol to raise the pH to biocompatible values and to eliminate the use of alcoholic solvents (19). The addition of biological molecules to a sol leads to their immobilization in the growing silica polymeric network as the fluid sol turns to a solid porous gel. According to the procedure initially devised in the authors' laboratories, the TMOS precursor is hydrolysed in water at pH ~ 2 to form the sol. Before adding the protein, a suitable buffer is mixed in to raise the pH to near biological values. A buffered solution of the protein is then added to the sol at optimal pH to prevent acid denaturation and/or solvent-induced aggregation of the protein. By using this method, a variety of proteins and enzymes have been immobilized within silica sol-gel glasses (1). The inorganic matrix makes the resulting solids chemically robust, thermally insensitive, mechanically well-defined, and dimensionally stable while the encapsulated biomolecule dopants lead to biocatalytic and/or ligand binding properties.

3.1 Sol-gel encapsulated myoglobin

Using the buffered method, it was possible to obtain optical quality silica monoliths, which made possible the characterization of the structure and function of the proteins by means of optical absorption spectroscopy. Myoglobin (Mb) was one of the first proteins to be encapsulated in the sol-gel materials by using the buffered sol-gel route (19). The effects of microencapsulation were monitored throughout the processes of gelation, aging, and drying of the gels. The optical absorption spectra throughout these processes are unchanged from that observed for the proteins in solution (19). The insensitivity of the band suggests that no structural and/or conformational changes occur during the sol-gel–xerogel transformation.

The solution chemistry of myoglobin can be reproduced in the sol-gel matrices. The as-prepared gels contain *met*myoglobin containing iron in the trivalent oxidation state. The oxidized *met*Mb can be reduced with sodium dithionite to deoxy Mb that contains iron in the +2 oxidation state. The reduction process can be monitored optically. The reduced form is stable as long as it is stored anaerobically. The deoxy form on exposure to air gives *oxy*Mb (MbO_2), and on treatment with CO gives carbonyl Mb (MbCO) (20). The reaction of the gaseous ligands with the sol-gel encapsulated protein are analogous to solution reactivity. An important aspect of this chemistry is that the spectroscopic changes associated with all of these reactions in the aged gels and the xerogels are similar to those which occur under the same conditions in solution. Thus, the xerogels obtained via the sol-gel method provide an elegant way of obtaining solid state materials of variable dimensions with biological properties.

Protocol 2.   Sol-gel encapsulation of myoglobin

Equipment and reagents
- Polystyrene semi-microcuvettes
- Parafilm
- TMOS sol
- Horse heart myoglobin (Sigma)
- 1 M ammonium hydroxide

Method

1. Dissolve 18 mg of myoglobin in 8 ml of deionized water.

2. Dilute 16 ml of freshly prepared TMOS sol with an equal volume of deionized water.

3. Add 266 µl of 1 M ammonium hydroxide to the diluted sol.

4. Add the solution of myoglobin prepared in step 1 to the sol mixture. Keep the mixture at low temperature in an ice-bath.

5. Pour 1 ml of the doped sol containing myoglobin in 1.5 ml polystyrene semi-microcuvettes.

6. Cover the cuvettes with Parafilm and allow the fluid sol to turn into a solid gel. It takes a few hours for the sol to gel.

7. Allow the covered sol-gels to age at ambient temperature for three days. The gels may be stored at 4°C. The as-prepared gels contain myoglobin in the *met* or Fe^{3+} form and must be reduced to the deoxy or Fe^{2+} form before it can bind dioxygen (for reduction to deoxy form see *Protocol 5*).

3.2 Cytochrome *c* in sol-gel thin films

A necessary requirement for thin film processing is that the initially formed sol should be stable for an extended period without becoming too viscous or forming a gel. Under controlled conditions, the hydrolysed precursors can be prevented from forming gels to yield stabilized fluid sols which are ideal for thin film deposition. The preparation of protein-doped sol-gel materials has TMOS precursor in buffers which raise the pH of the medium to near biological values ($\sim$ 7). Conditions of high pH lead to short gelation times and the sol becomes macroscopically rigid within minutes. Such gelation conditions are incompatible with the common sol-gel film deposition approaches where thin films are formed from a low viscosity sol. If protein-doped thin films are to be synthesized, longer gelation times are necessary. Inasmuch as sol-gel polymerization is a two-step process involving hydrolysis and condensation of the hydrolysis products, chemical strategies that either reduce the concentration of the hydrolysed silicate precursor or slow down condensation ought to result in stabilized sols with longer gelation times. A lower water to TMOS ratio and high methanol concentration conditions retard the primary hydrolytic step and are conducive for forming stable sols with considerably longer gelation times. Another way to achieve longer gelation times is to lower the pH. The higher proton concentration in the medium inhibits the formation of water-derived oxo-bridges in the polycondensation step, thus yielding a stabilized sol. Unfortunately, increased concentrations of alcohol and lower pH are detrimental to protein stability.

In the synthesis method devised in the authors' laboratories, a combination of both strategies was used to obtain longer gelation times without compromising the stability of the protein. One composition which combined long gelation time and protein stability, and produced high quality optical thin films was the ratio 40:50:10 (vol. percentages) of MeOH, TMOS sol, and buffered solution of cytochrome *c*, respectively. Dip coating techniques can be used to produce films. Excellent quality protein-doped thin films that were homogeneous, optically transparent, crack-free, and adherent to the substrate (silicon or glass microscope slides) were produced by this method (21).

Optical absorption characteristics of the haem Soret band of the ferricytochrome *c* are able to establish the structural integrity of the sol-gel trapped protein. The absorption spectrum for a reference solution of the protein dissolved in pH 4.25 acetate buffer (0.1 M) has its Soret maximum at 407 nm, whereas the absorption maximum of the as-prepared film is centred at 405 nm

(*Figure 5a*). Immersing the thin films in either a 0.1 M acetate buffer (pH 4.25) or 0.1 M phosphate buffer (pH 7) does not alter the optical absorption characteristics of the films. The protein-doped thin films also exhibit the characteristic redox properties of ferricytochrome *c*. Immersing the as-prepared films in a solution of sodium dithionite results in an increased intensity peak in the ~ 550 nm region, which is indicative of the reduced form of the protein. The reduction of the encapsulated protein is also accompanied by a 4 nm red shift of the Soret transition to 409 nm and a concomitant change in intensity (*Figure 5b*). Air oxidation of the reduced form produces the oxidized species and the original spectrum could be re-obtained. It is important to emphasize that the redox changes and accompanying optical absorption changes occurring in these films are totally reversible, and the redox forms of the protein can be successively interconverted without any significant deterioration in the

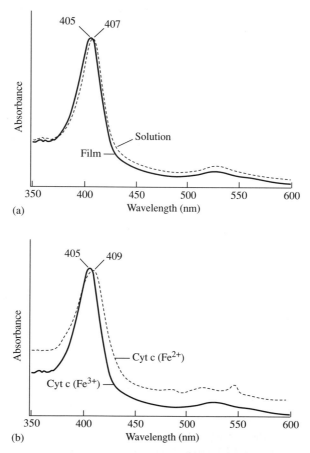

Figure 5. Optical absorption spectra of cyt *c* immobilized in sol-gel thin films. (a) Comparison of solution and as-prepared thin films containing immobilized cyt *c*. (b) Spectral changes associated with dithionite reduction of cyt *c* in as-prepared films.

optical properties of the films. These results are in general agreement with redox-dependent spectral changes observed in aqueous media and in monolithic sol-gels. The protein-doped thin films respond much more rapidly to the external reagents possibly due to smaller diffusional distances. Remarkably, total reduction of the protein in these thin films was observed after only a few minutes of immersion with sodium dithionite as compared to monolithic glass samples (1 cm × 1 cm × 2 cm) where about an hour is needed for a similar change to take place.

Protocol 3. Dip coated sol-gel thin films containing immobilized cytochrome *c*

Equipment and reagents
- Glass microscope slides (3 cm × 1 cm × 1 mm)
- Petri dishes
- Dip coating film puller
- No-Chromix
- Freshly prepared TMOS sol
- Horse heart cytochrome *c* (Sigma, Type VI)
- 0.1 M sodium acetate buffer pH 4.25
- 1 M hydrochloric acid
- Methyl alcohol

Method
1. Prepare three baths; one containing No-Chromix and two containing deionized water.
2. Clean glass slide substrates by immersing first in the No-Chromix bath for 1 min, followed by sequential immersion in the two water-baths each for 1 min.
3. Rinse the substrates under running deionized water.
4. Store the substrates in a beaker in deionized water.
5. Dry the substrates (glass slides) by blowing dry nitrogen or argon gas over them.
6. Attach the substrates to the free end of the Mylar tape in the film puller.
7. Prepare 0.4 ml of cytochrome *c* solution by dissolving 50 mg of the protein in 0.3 ml of 0.1 M acetate buffer and 0.1 ml of 1 M HCl. Keep the solution in an ice-bath.
8. Mix 2 ml of freshly prepared TMOS sol with 1.6 ml of methanol, and to this mixture add 0.4 ml of the buffered cytochrome *c* solution. Store the mixture at low temperature in an ice-bath or in a refrigerator.
9. Pour the mixture into the container for film pulling and withdraw the substrates at a slow rate (~ 10 cm/min) by monitoring the water flow through the control valve.
10. Allow the freshly formed films to dry at room temperature in a covered Petri dish for about 10–15 min. Uniformly deposited thin films are characterized by an iridescent surface. Store the films in a refrigerator.

---

4. Sol-gel encapsulation of enzymes

The encapsulation of simple enzymes and coupled systems has also been accomplished in SiO_2-based sol-gel media. The enzyme systems that have been immobilized in sol-gels are oxidases and dehydrogenases. Several oxidases such as glucose oxidase, oxalate oxidase, and dehydrogenases such as alcohol dehydrogenase, and glucose-6-phosphate dehydrogenase have been immobilized in sol-gel media. The coupled systems use the oxidases in combination with horseradish peroxidase (HRP) to perform a series of reactions; the catalysis products of one enzyme (oxidase) serve as reactants for the other enzyme (HRP). The fact that such coupled reactions do take place in the porous matrices demonstrates the ability to use sol-gel matrices to devise enzymatic systems capable of carrying out multistep catalytic transformations.

Biocatalysis is sensitive to microenvironmental effects. The effects of sol-gel immobilization upon the reactivity of the enzyme can be estimated by using Michaelis–Menten kinetics. A generalized reaction for catalysis by an enzyme is represented as follows:

$$E + S \underset{k_{-1}}{\overset{k_1}{\rightleftharpoons}} ES \overset{k_2}{\longrightarrow} E + P$$

where E is the enzyme, S is the substrate, and P is the product of the reaction. The Michaelis constant K_m ($\sim k_{-1}/k_1$) measures the dissociation of the enzyme:substrate complex and in turn serves as an estimate of its stability. An increased value of the Michaelis constant implies that the equilibrium is shifted towards the free enzyme and substrate, and suggests a relatively weaker enzyme:substrate complex. Another parameter, k_{cat} ($\sim k_2$), called the turnover number, estimates the rate of formation of the product from the enzyme:substrate complex. The ratio k_{cat}/K_m, on the other hand, represents the *apparent* rate constant for combination of a substrate with the free enzyme.

4.1 Sol-gel encapsulated glucose oxidase

Glucose oxidase (GOx) is one of the enzyme systems that was investigated for encapsulation in transparent sol-gel glasses (22). The enzyme catalyses the oxidation of glucose to gluconolactone using oxygen. An important requirement in preservation of the catalytic activity is that there must not be a substantial structural change during the process of matrix confinement. In order to quantify the enzyme activity, the turnover number (k_{cat}) and the apparent dissociation constant (K_m) with β-D-glucose is determined. The k_{cat} for encapsulated glucose oxidase (250 sec^{-1}) is identical to that found for the enzyme in solution (251 sec^{-1}). The K_m value in the sol-gel medium (0.05 M) for the glucose:glucose oxidase complex (glu-GOx) is about twice that found in the solution (0.028 M) suggesting either a decreased tendency for the association reaction or an increased aptitude for the enzyme:substrate complex to dissociate into the products. However, the unaltered k_{cat} value suggests that the rate of product formation is not changed as a result of immobilization. Thus, it can be concluded that in the gel the formation constant for the glu-GOx complex is reduced. This may be due to the reaction rate becoming limited by mass transport, that is, the substrate concentration in the gel is reduced compared to solution because of depletion through enzyme turnover. A comparison of the apparent association constant (k_{cat}/K_m) for glu-GOx system in the solution (~ 9000 M^{-1} sec^{-1}) and in the gel (5000 M^{-1} sec^{-1}) shows that the rate of associative enzyme substrate complex formation is reduced by half in the gel matrix.

Protocol 4. Sol-gel encapsulation of glucose oxidase

Equipment and reagents
- Polystyrene cuvettes
- Parafilm
- TMOS sol
- Glucose oxidase (Sigma, Type X-S from *Aspergillus niger*)
- 0.01 M sodium phosphate buffer pH 6

Method
1. Dissolve 4.38 mg of glucose oxidase in 3 ml of 0.01 M phosphate buffer. Store at 4°C.
2. Adjust the pH of freshly prepared TMOS sol (4 ml) to ~ 6 by adding 0.01 M phosphate buffer (~ 5 ml). Keep the sol in an ice-bath.
3. To this sol, add 3 ml of the solution containing glucose oxidase.
4. Transfer the sol quickly to polystyrene cuvettes for casting of gels.
5. Cover the cuvette with Parafilm and allow the fluid sol to turn into a solid gel. The gelation time is approx. 2–3 min.
6. Allow the covered sol-gels to stand in a refrigerator at 4°C for two to three days to age. The transparent sol-gels will shrink and separate from the cuvette during this time with expulsion of clear solvent liquid.
7. Make two or three pin-holes in the Parafilm to dry the sol-gels. It usually takes three to four weeks for the sol-gels to dry at 4°C.

4.2 Sol-gel encapsulated oxalate oxidase

Oxalate oxidase catalyses the oxidation of oxalate to carbon dioxide. Immobilization in the sol-gel matrix produces solid state materials containing the reactive enzyme (18). The reactivity of the enzyme was probed by kinetics studies. The value of the Michaelis constant (K_m) increases for the sol-gel immobilized system ($K_m = 1.1 \times 10^{-4}$ in solution and 4.1×10^{-4} in aged gel). The approximately fourfold increase indicates that the binding of the oxalate ($C_2O_4^{2-}$) with the enzyme is weaker in the glass. The apparent association constant

(k_{cat}/K_m) for the oxalate:oxalate oxidase complex is substantially altered as a result of confinement of the enzyme in the sol-gel matrix (k_{cat}/K_m = 170 in solution and 2.3 in aged gel). This implies a relatively destabilized enzyme: substrate complex. Additionally, the k_{cat} parameter is also reduced in the gel ($k_{cat}/10^{-4}$ = 187 in solution and 9.4 in aged gel), suggesting deactivation of the product forming step in the gel medium. In this case, while the dissociative product forming step is reduced by a factor of 20 (187/9.4), the associative step is reduced by a factor of 75 (170/2.3).

The reactivity of the immobilized oxalate was also tested by using a mono-protonated oxalate ion ($HC_2O_4^-$). In a pH 3 solution, approximately 90% of the substrate exists in the monoprotonated form. Reaction rates can be determined using dianion (at pH 6) and monoanion (at pH 3). The sol-gel material itself has an isoelectric point ~ 2 and above this pH it is negatively charged. Based on electrostatic arguments it can be expected that diffusion of the dianion into the matrix will be impeded much more than that of monoanion. Experimental rate data, however, shows that the ratio of the rates at the two different pHs between the enzyme in solution (3.8 ± 0.2) and the sol-gel (3.7 ± 0.2) is almost identical. The results prove that:

(a) The diffusion of the substrates into the matrix (at least for the oxalate system) is not very important.

(b) The pore sizes of the matrix are sufficiently larger than the substrates such that, in general, the matrix does not significantly interact with the smaller substrates.

5. Sol-gel-based biosensor elements

The sol-gel glasses are transparent and therefore provide an ideal matrix for development of optical sensor elements based on the reactivity of immobilized enzymes. The high specificity and rates of reaction make enzymes excellent reagents for analytical detection and biosensing. The encapsulation of biomolecules in sol-gel matrices provides an efficient biosensing design where the motion of the recognition molecule is restricted while the flow of the analytes is allowed through the porous structure. The optical transparency is especially conducive for optical monitoring of reaction events using both absorption as well as emission spectroscopy (23). The porous structure of the sol-gels allows diffusional migration of the external analyte molecules to the reactive sites of the enzymes. The reaction between the encapsulated enzyme and the molecule is used to initiate an optical response which is directly correlated with the concentration of the enzyme. Using sol-gel materials, different types of optical sensor elements (*Figure 6*) have been designed where the optical signal is generated as a result of:

- ligand binding to a metalloenzyme
- formation of a coloured dye

Biosensor based on metalloenzyme-ligand binding

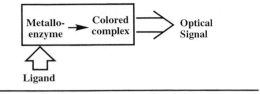

Biosensor based on dye formation

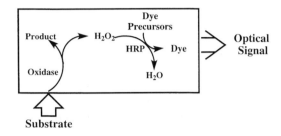

Figure 6. Schematics of sol-gel-based optical biosensors.

Each of the sensor elements was tested with monolithic samples. For a biosensor element based on coloured metalloenzymes, the changes in absorption spectra are correlated with the concentration of the ligand, and O_2, CO, and NO sensors are based on absorption spectral changes of the haem group. However, with enzymes that do not absorb in the visible region, an alternative strategy was employed, wherein the enzyme:substrate reaction is coupled with the generation of an optically absorbing dye from its precursors. The glucose and oxalate sensors using immobilized oxidase/peroxidase enzymes employ this method to generate an optical signal.

For enzymes acting as centres of exogenous ligand binding, the resultant changes in optical properties can be utilized for detection purposes. On the other hand, for the enzymes involved in catalysis, the changes associated with either a decrease of reactant or an increase of product concentrations serve as useful monitors. Thus far, biogels have been successfully utilized for detection of dioxygen, glucose, and oxalate. The transparent gel matrix allows optical transduction of the signal. The unique features of the sol-gel-derived SiO_2 glasses are their optical transparency to 300 nm and sufficiently small pore sizes so that there is minimal scattering of light. These properties make these materials suitable for sensor design based on optical methods of transduction.

5.1 Biosensor element for dissolved oxygen

Myoglobin is a haem protein that functionalizes binding of atmospheric dioxygen. The high affinity of this protein for O_2 coupled with the changes in visible absorption spectra provide an opportunity to develop a dioxygen sensor. In order to quantify the optical changes for applications as biosensor elements, myoglobin is chosen as a model system (24). Due to the very fast reaction with gaseous oxygen, experiments are done with dissolved oxygen (DO).

Protocol 5. Preparation of deoxymyoglobin sol-gels

Equipment and reagents
- Polystyrene semi-microcuvettes
- Erlenmeyer flask
- Spectrophotometer
- Syringe with needle
- 0.01 M Tris buffer pH 7.4
- Aged silica sol-gels containing myoglobin (*Protocol 2*)
- Sodium dithionite
- Degassed water
- Argon gas

Method
1. Dissolve 100 mg of sodium dithionite in 300 ml of 0.01 M Tris buffer in an Erlenmeyer flask.
2. Immerse aged sol-gels containing encapsulated *met*myoglobin (prepared using *Protocol 2*) in the dithionite solution. Bubble argon gas continuously through the solution to prevent air oxidation.
3. Allow the reaction to go to completion at ambient temperature for about 2 h while the argon gas is continuously bubbled through the solution. During this time the colour of the gels changes from pale green to a bright brownish yellow. These reduced gels can be used for detection of oxygen. The reduced sol-gels may be stored in the dithionite solution under argon atmosphere at 4°C.

The initial objective is to establish a correlation of the absorption spectral characteristics with DO concentration. The prominent differences in absorption maxima of the Soret transition of the deoxy and *oxy* form in the 430 nm region allow a convenient way to quantify protein:dioxygen interactions (*Figure 7*). By monitoring the decrease in intensity of the 436 nm transition of the deoxy form, a linear relationship is established between the optical response of Mb and DO concentration (*Figure 7*, inset). Moreover, only one wavelength needs to be monitored to determine the DO concentration. The analytical advantages of this novel biosensing element are linear correlation, short response intervals ($\sim$ 1 min), high specificity, reversibility of the *oxy* form to deoxy state and thus, cost-effectiveness in terms of reusability of the sensing biogel element. Apart from the utility of Mb biogels as O_2 sensors, they can also be used as a sensor for CO by taking advantage of the binding of CO and monitoring the distinct changes occurring in the absorption spectrum (20).

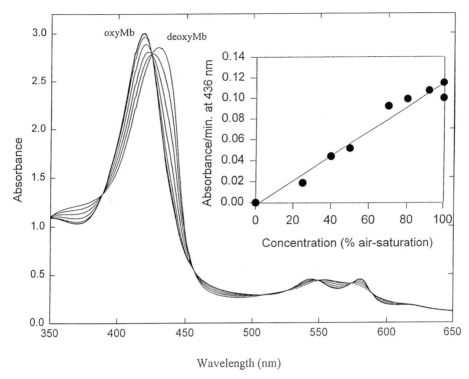

Figure 7. Changes in optical absorption spectra shown by the deoxyMb containing sol-gels upon treatment with dissolved oxygen. Inset shows the correlation between the rate of absorption change at 436 nm with dissolved oxygen.

Protocol 6. Preparation of sol-gel-based oxygen sensor elements

Equipment and reagents

- Polystyrene cuvettes
- Membrane equipped oxygen meter
- Spectrophotometer
- Syringes with needle

- Silica sol-gels containing deoxymyoglobin (*Protocol 5*)
- Argon gas

Method

1. Prepare degassed water by continuously bubbling argon through it for about 2 h at ~ 60°C.

2. Transfer the sol-gels containing deoxymyoglobin (*Protocol 5*) into a cuvette (standard 4.5 ml total volume) along with dithionite solution. Quickly cover the cuvettes with a Parafilm seal.

3. Remove excess dithionite solution from the cuvette using a syringe with needle.

131

Protocol 6. *Continued*

4. Introduce 3.5 ml degassed water into the cuvette using another clean syringe.

5. Remove the wash water from the cuvette after 30 min.

6. Repeat steps 4 and 5 three times to remove any unreacted dithionite from the gel samples.

7. Introduce unknown water sample whose dissolved oxygen concentration is to be determined into the cuvette with the aid of a microsyringe.

8. Record optical density of the gels at 436 nm immediately after the introduction of unknown at 1 min intervals for 5 min using a spectrophotometer.

9. Determine the absorbance rate change. Unknown concentrations may be determined using a calibration curve based on known concentrations of dissolved oxygen (using an oxygen meter).

5.2 Biosensor element for glucose

The monitoring of blood glucose levels has many clinical applications. The biosensors for detection of glucose utilize the enzyme glucose oxidase (GOx), which catalyses air oxidation of β-D-glucose to give gluconic acid and hydrogen peroxide. Using the biogel methodology, a glucose sensor element is developed using immobilized GOx in combination with horse-radish peroxidase (HRP) and suitable dye precursors, to take advantage of the catalysis induced by peroxidase between the dye precursors and hydrogen peroxide. The dye precursors used are 4-aminoantipyrine and *p*-hydroxybenzene sulfonate which form a quinoneimine dye with an absorption maximum at 510 nm. On exposure of gel monoliths containing the simultaneously immobilized GOx and HRP to glucose, the GOx catalyses generation of hydrogen peroxide, and in turn, triggers the second reaction, i.e. the formation of dye from its precursors catalysed by HRP. The rate of formation of coloured dye is directly proportional to the concentration of the glucose present (22). By monitoring the absorption maximum of the dye generated, the amount of glucose in the surrounding solution can be determined.

Protocol 7. Preparation of sol-gel-based glucose sensor element

Equipment and reagents

- Polystyrene cuvettes
- Parafilm
- TMOS sol
- Glucose oxidase (Sigma, Type X-S from *Aspergillus niger*)

- Horse-radish peroxidase (Sigma, Type II)
- 4-aminoantipyrine
- *p*-Hydroxybenzene sulfonate
- 0.01 M sodium phosphate buffer pH 6

Method

1. Prepare stock solutions of glucose oxidase (1.46 mg/ml) and horse-radish peroxidase (0.96 mg/ml) using 0.01 M phosphate buffer pH 6.

2. Adjust the pH of freshly prepared TMOS sol (4 ml) to ~ 6 by adding 0.01 M phosphate buffer (~ 5 ml). Keep the sol in an ice-bath.

3. To this sol, add 3 ml of enzyme solution that is made by adding 21 μl of glucose oxidase stock solution, 138 μl of peroxidase stock solution, and 2.84 ml of 0.01 M phosphate buffer pH 6.

4. To the resulting mixture, add 8.32 ml of the dye precursor solution containing 0.5 mM of 4-aminoantipyrine and 2 mM *p*-hydroxybenzene sulfonate dissolved the phosphate buffer.

5. Transfer the sol quickly to polystyrene cuvettes for casting of gels.

6. Cover the cuvette with Parafilm and allow the fluid sol to turn into a solid gel. The gelation time is approx. 2–3 min.

7. Allow the covered sol-gels to stand in a refrigerator at 4°C for two to three days to age. Reaction of these gels with glucose results in formation dye with an absorption maximum at 520 nm. These gels can be used as sensor elements to detect unknown concentrations of glucose using a calibration curve.

6. Concluding remarks

While there exist a number of immobilization strategies, the sol-gel approach is quite unique due to its simplicity and its provision of an optically transparent material. Initial studies have established a general applicability of the approach for immobilization of different proteins and enzymes. In all the cases investigated to date, the structure and reactivity of the biological molecules are retained upon immobilization in sol-gel media. We have illustrated that the materials containing encapsulated enzymes are biocatalytic and have many areas of applications including optical biosensors. The success of these systems suggests that a whole range of novel biocatalytic materials, optical biosensors, and biomedical devices may be fabricated via a judicious choice of the dopant biomolecule.

Acknowledgements

We express our appreciation to K. E. Chung, L. M. Ellerby, E. H. Lan, J. M. Miller, C. R. Nishida, F. Nishida, H. M. Soyez, and S. A. Yamanaka for their contributions and for their involvement with various aspects of the work. We are also grateful to the National Science Foundation for their financial support of this work.

References

1. Dave, B. C., Dunn, B., Valentine, J. S., and Zink, J. I. (1994). *Anal. Chem.*, **66**, 1120A.
2. Brinker, C. J. and Scherer, G. (1989). *Sol-gel science: the physics and chemistry of sol-gel processing*. Academic Press, San Diego.
3. Hench, L. L. and West, J. K. (1990). *Chem. Rev.*, **90,** 33.
4. Dave, B. C., Dunn, B., and Zink, J. I. (1995). In *Access in nanoporous materials* (ed. T. J. Pinnavaia, and M. F. Thorpe), p. 141. Plenum Press, New York.
5. Dunn, B. and Zink, J. I. (1991). *J. Mater. Chem.*, **1**, 903.
6. Dave, B. C., Dunn, B., Valentine, J. S., and Zink, J. I. (1996). In *Nanotechnology: molecularly designed materials* (ed. G.-M. Chow and K. E. Gonsalves), p. 351. American Chemical Society, Washington DC.
7. Livage, J., Henry M., and Sanchez, C. (1988). *Prog. Solid State Chem.*, **18** 259.
8. Iler, R. K. (1979). *The chemistry of silica*. John Wiley, New York.
9. Brinker, C. J., Drotning, W. D., and Scherer, G. W. (1984). In *Better ceramics through chemistry* (ed. C. J. Brinker, D. E. Clark, and D. R. Ulrich), p. 25. Elsevier, North Holland, New York.
10. Scherer, G. W. (1989). *J. Non-Cryst. Solids*, **109**, 183.
11. Klein, L. C. (ed.) (1988). *Sol-gel technology for thin films, fibers, preforms, electronics, and specialty shapes*. Noyes Publications, New Jersey.
12. Schmidt, H. (1992). *Struct. Bonding*, **77**, 119.
13. Klein, L. C. (ed.) (1994). *Sol-gel optics: processing and applications*. Kluwer Academic Press, Boston.
14. Brinker, C. J., Hurd, A. J., Frye, G. C., Ward, K. J., and Ashley, C. S. (1990). *J. Non-Cryst. Solids*, **121**, 294.
15. Nishida, F., McKiernan, J. M., Dunn, B., Zink, J. I., Brinker, C. J., and Hurd, A. J. (1995). *J. Am. Ceram. Soc.*, **78**, 1640.
16. Brinker, C. J., Hurd, A. J., Schunk, P. R., and Ashley, C. S. (1991). *J. Ceram. Soc. Jpn.*, **99**, 862.
17. Avnir, D., Braun, S., Lev, O., and Ottolenghi, M. (1994). *Chem. Mater.*, **6**, 1605.
18. Zink, J. I., Valentine, J. S., and Dunn, B. (1994). *N. J. Chem.*, **18**, 1109.
19. Ellerby, L. M., Nishida, C. R., Nishida, F., Yamanaka, S. A., Dunn, B. S., Valentine, J. S., *et al.* (1992). *Science*, **255**, 1113.
20. Lan, E. H., Davidson, M. S., Ellerby, L. M., Dunn, B., Valentine, J. S., and Zink, J. I. (1994). *MRS Symp. Proc.*, **330**, 289.
21. Dave, B. C., Soyez, H., Miller, J. M., Dunn, B., Valentine, J. S., and Zink, J. I. (1995). *Chem. Mater.*, **7**, 1431.
22. Yamanaka, S. A., Nishida, F., Ellerby, L. M., Nishida, C. R., Dunn, B., Valentine, J. S., *et al.* (1992). *Chem. Mater.*, **4**, 495.
23. Yamanaka, S. A., Dunn, B., Valentine, J. S., and Zink, J. I. (1995). *J. Am. Chem. Soc.*, **117**, 9095.
24. Chung, K. E., Lan, E. H., Davidson, M. H., Dunn, B. S., Valentine, J. S., and Zink, J. I. (1995). *Anal. Chem.*, **67**, 1505.

8

Immobilization of 'smart' polymer–protein conjugates

PATRICK S. STAYTON and ALLAN S. HOFFMAN

1. Introduction

The remarkable molecular recognition capabilities of proteins provide the basis for many important clinical, industrial, and laboratory applications. Key existing technologies include separations, biosensors, biotransformations, diagnostics, and targeted drug delivery. The central event in protein-mediated recognition processes is typically a protein:ligand binding step, which is characterized by a combination of high specificity and affinity, and in the case of enzymes by a high affinity for the reaction coordinate transition state. While these molecular recognition capabilities have proven extremely useful, many practical problems limit the utility of proteins because nature has not optimized them for the device environment. Serious problems with isolated proteins include:

(a) Low stability—tendency to denature.

(b) Susceptibility to degradation by proteases, bacteria, etc.

(c) Need for harsh elution or regeneration conditions because of the tight binding to targets.

As a result, some key challenges of general importance to technology development include the development of immobilization strategies that:

(a) Optimize protein stability and activity.

(b) Provide controlled mass transport of the target, cofactors, contaminants, etc. through the device matrix.

(c) Provide precise control of target capture and/or release.

In this chapter, we will describe and discuss how the unique properties of stimuli-responsive polymers can be used to provide new routes for addressing these challenges. Polymers which respond with large physical changes to small changes in environmental conditions are called stimuli-responsive, environmentally-sensitive, 'intelligent', or 'smart' polymers (1, 2). These polymers display a reversible change in size and cycle between a water soluble, extended random coil conformation and a phase separated, collapsed conformation in response to mild changes in:

(a) Physical stimuli:

- temperature
- solvents
- electromagnetic radiation
- electric fields
- mechanical stress, strain
- sonic radiation
- magnetic fields

(b) Chemical stimuli:

- pH
- specific ions
- ionic strength
- biochemical stimuli
- enzyme substrates
- affinity ligands

2. Applications of stimuli-responsive polymers and gels

2.1 Stimuli-responsive hydrogels

Enzymes can be immobilized in hydrogels and their activity controlled by the reversible collapse and swelling that is triggered by signals such as pH, temperature, specific ions, etc. The mass transport of substrates, products, cofactors, etc. can also be reversibly controlled in and out of the gel by the cyclic swelling and collapse. This type of activity allows for the controlled release of drugs from loaded hydrogel carriers, where for example the change in pH between the gastric and enteric compartments can trigger the swelling and delivery of drugs at controlled rates. Surface coatings on membranes or other surfaces can be similarly used as permeation switches that are again reversibly controlled by environmental signals. *Figure 1* illustrates some of these applications.

2.2 Stimuli-responsive polymer–protein conjugates

The conjugation of soluble stimuli-responsive polymers to proteins was first carried out by Monji and Hoffman in 1986 (3). We applied the poly-*N*-isopropylacrylamide–monoclonal antibody (PNIPAAm-MAb) conjugate to a temperature-induced phase separation immunoassay (3). In subsequent studies, similar soluble–insoluble conjugates with protein A, and with enzymes such as

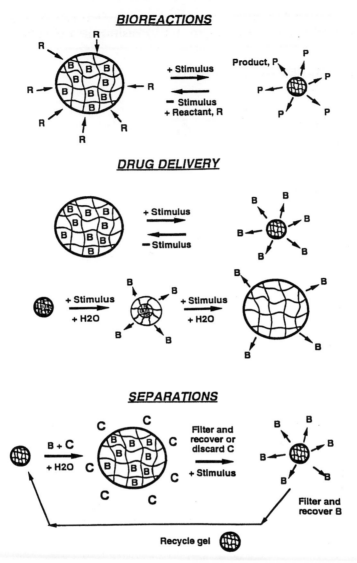

Figure 1. Some applications of stimulus-responsive hydrogels in bioreactors, drug delivery and separations.

β-galactosidase, asparaginase, and trypsin were prepared. We applied these conjugates for temperature-induced affinity separations (protein A–IgG) and enzyme–product separation, recovery, and enzyme recycle (4–7). We have also conjugated other biomolecules to PNIPAAm, including RGD peptide sequences (8), and phospholipids (9). *Figure 2* illustrates some of these schematic applications of these conjugates.

Uses of Intelligent Polymer-Biomolecule Conjugates

(1) ENZYME RECOVERY AND RECYCLE

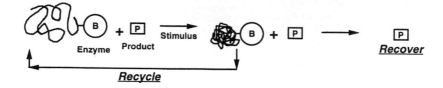

(2) LIGAND RECEPTOR RECOVERY AND/OR REMOVAL

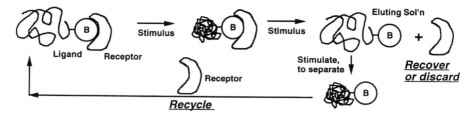

(3) ASSAY OF ANALYTE IN A COMPLEX MIXTURE

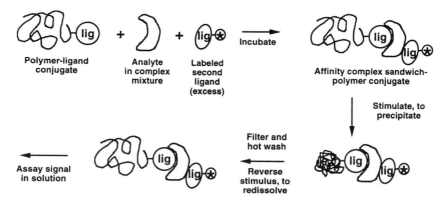

Figure 2. Applications of stimulus-responsive polymer–protein conjugates in protein recovery and analysis.

2.3 Site-specific stimuli-responsive polymer–protein conjugates

We have recently reported the site-specific conjugation of an end-reactive stimuli-responsive polymer to a genetically engineered protein (10). This conjugate was prepared in order to precisely control the polymer:protein stoichiometry and also to locate the polymer away from the protein active site, so that a physical phase separation could be carried out without interfering with the recognition site binding activity (*Figure 3*). Conventional protein–polymer conjugation schemes use lysine amino groups, but this approach does not easily provide control over the location or number of attachment sites. Protein engineering techniques have allowed us to design unique attachment sites on the protein surface for polymer conjugation. The synthesis of a polymer with one activated group per polymer chain, and at one end of the molecule, then allows the stoichiometrically precise attachment of the polymer to the protein of interest.

This approach has recently been used to open a new class of polymer–protein applications, where the stimuli-responsive polymer is used as a molecular switch to control protein activity (*Figure 4*). For example, a genetically engineered cysteine residue has been introduced into streptavidin and subsequently conjugated to a temperature-sensitive polymer (11). The reversible, thermally-induced collapse of the polymer then acts as a 'molecular gate' to control the association of biotin with streptavidin, which is bound on the surface of a microporous membrane. The environmentally triggered 'gating'

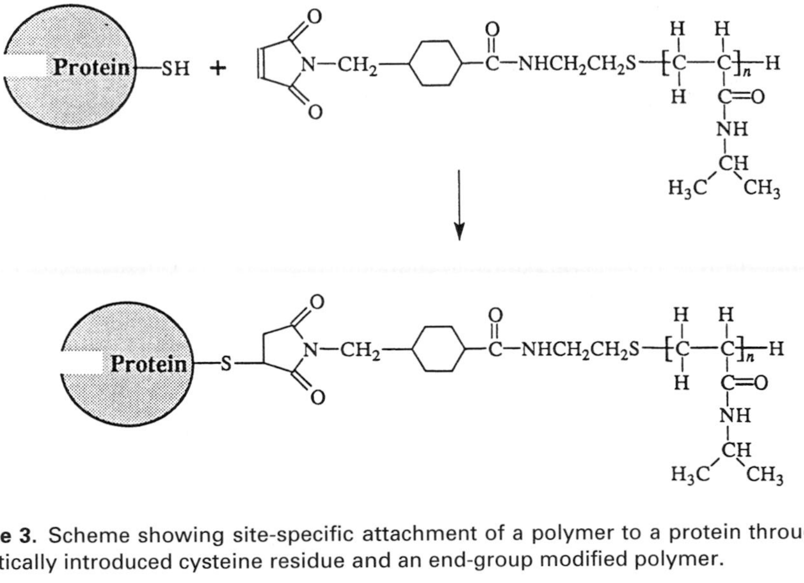

Figure 3. Scheme showing site-specific attachment of a polymer to a protein through a genetically introduced cysteine residue and an end-group modified polymer.

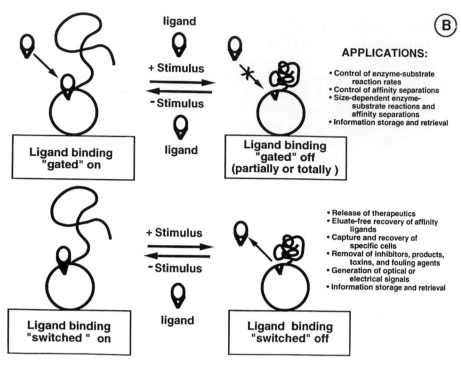

Figure 4. Use of stimulus-reponsive polymers as molecular switches to control the activities of proteins.

and 'switching' capabilities of these 'intelligent' polymer–protein conjugates have many potential uses that include control of:

- biorecognition processes in general
- enzyme:substrate size selectivity and reaction rates
- affinity ligand size selectivity and binding rates
- generation of optical or electrical signals
- eluate-free recovery of affinity ligands
- eluate-free recovery of specific cells
- release of site bound biological or chemical agents

3. Synthesis of stimuli-responsive polymers and gels

3.1 Temperature-sensitive PNIPAAm

Protocol 1. Synthesis of PNIPAAm

Reagents
- *N*-isopropylacrylamide (NIPAAm) (Eastman Kodak)
- 2,2'-azoisobutyronitrile (AIBN) (J. T. Baker)

Method

1. The NIPAAm was purified by recrystallization fron *n*-hexane and dried in vacuum.
2. AIBN was recrystallized from methanol.
3. PNIPAAm is synthesized by the radical polymerization of NIPAAm using AIBN as initiator.
4. 0.1 mole of NIPAAm and AIBN (0.2% by weight) in 40 ml alcohol solvent are placed in a thick-walled polymerization tube, and the mixture degassed by freezing and evacuating and then thawing four times.
5. After cooling for the last time, the tubes are evacuated and sealed prior to polymerization. Tubes are then immersed in a water-bath at 50°C for 24 h.
6. The resulting polymer is isolated by precipitation into 2 litres diethyl ether, collected by filtration, and weighed to determine yield.

3.2 Temperature- and pH-sensitive polymers and gels

Heterogeneous hydrogels have been synthesized that have both thermal and pH responsiveness (12). The polymers are prepared from NIPAAm, acrylic acid (AAc), and vinyl terminated polydimethylsiloxane (VTPDMS). Other groups have used copolymers of hydroxyethyl methacrylate and methacrylic acid or maleic anhydride to synthesize pH-sensitive hydrogels (13). Thermal- and pH-sensitive polymers have been prepared that can be conjugated to proteins. For example, co-polymerization of NIPAAm with 4-pentenoic acid using AIBN as an initiator proceeds at 50°C with total monomer concentration of 5 mole/litre.

3.3 Thiol-reactive polymers

Protocol 2. Maleimide terminated polymers

Reagents

- *N*-Isopropylacrylamide
- (NIPAAm) (Eastman Kodak)
- 1-Aminoethanethiol-hydrochloride   or   β-mercaptopropionic acid

- 2,2'-Azoisobutyronitrile (AIBN) (J. T. Baker)
- Succinimidyl-4-(*N*-maleimidomethyl) cyclohexane-1-carboxylate

Method

1. The synthesis of an amino terminated polymer proceeds by the radical polymerization of NIPAAm in the presence of AIBN as an initiator and 1-aminoethanethiol–hydrochloride as a chain transfer reagent.

2. To synthesize a chain with –COOH or –OH terminal groups, carboxyl- or hydroxyl-thiol chain transfer agents are used, respectively, instead of the amino-thiol. For example, carboxylated PNIPAAm is synthesized using β-mercaptopropionic acid (Aldrich). It should be noted here that the synthesis of the end-reactive polymers is based on a chain transfer initiation and termination mechanism, and this yields a relatively short polymer chain, having a M_r somewhere between 1000 and 25–30 000. The shortest chains, less than 10 000 in M_r, are usually called 'oligomers'. Oligomers of different molecular weights can be synthesized by simply changing the ratio of monomer to chain transfer reagent.

3. The maleimide terminated oligo(NIPAAm) [MI-oligo(NIPAAm)] is synthesized by reacting the amino end-group with succinimidyl-4-(*N*-maleimidomethyl)cyclohexane-1-carboxylate, resulting in MI-oligo(NIPAAm).

Protocol 3. Vinyl sulfone (VS) terminated polymers

Reagents

- Hydroxyl terminated poly (NIPAAm)
- Divinylsulfone

- Triethylamine
- Diethylether

Method

1. NIPAAm-VS can be prepared by reacting a 5 mole % excess of divinyl sulfone and 5 mole % excess of triethylamine with hydroxyl terminated poly(NIPAAm) ($M_n = 3800$) in chloroform for 24 h at room temperature.

2. The product is recovered by precipitation with 2 litres ethyl ether.

3. The product should show characteristic vinyl sulfone peaks in the proton NMR spectrum (in d_6-DMSO) at 6.21 p.p.m. (two hydrogens) and 6.97 p.p.m. (one hydrogen).

4. Conjugation of stimuli-responsive polymers to proteins

4.1 Conjugation to protein amino groups

Protocol 4. Conjugation to protein amino groups (*Figure 5*)

Reagents
- 0.3 M Sodium phosphate pH 6.8
- N-Isopropylacrylamide (NIPAAm)
- N-Acryloxysuccinimide
- 2,2′-Azoisobutyronitrile (AIBN)
- Hydroxlapatite
- Tetrahydrofuran (THF)
- Toluene
- Petroleum ether
- Dimethylformamide (DMF)
- 0.01 M Sodium phosphate pH 9
- Saturated ammonium sulfate

Method
1. Activated PNIPAAM was made by co-polymerizing NIPAAm with N-acryloxysuccinimide (NAS) (1–10 mole % depending on degree of desired reactive groups) in anhydrous THF/toluene using AIBN as initiator. 0.1 mole NIPAAM and 1–10 mole % NAS are dissolved in the mixture of THF and toluene with nitrogen bubbled through the solution. AIBN is added at 0.2% weight and reacted at 50°C for 24 h with continuous stirring. Petroleum ether is added to precipitate the activated copolymer. The polymer is filtered, washed with petroleum ether, and dried in vacuum.
2. 30 mg of activated PNIPAAm in 100 µl DMF is added to 2 mg protein in 1 ml 0.01 M sodium phosphate buffer pH 9.
3. After thorough mixing, the solution is incubated at room temperature for 4 h.
4. The volume is adjusted to 3 ml with deionized water and 0.5 ml of saturated ammonium sulfate added to precipitate the unconjugated polymer and the polymer–protein conjugate.
5. The precipitate is collected by centrifugation at 15 000 r.p.m. at 20°C for 30 min.
6. The precipitate is redissoved in 3 ml of 0.01 M sodium phosphate buffer pH 9 and reprecipitated by adding 0.5 ml of saturated ammonium sulfate.
7. The final precipitate is dissolved in 6 ml deionized water and applied to a hydroxylapatite column (1.6 × 1.6 cm) which is equilibrated with deionized water.

Protocol 4. *Continued*

8. pNIPAAm is eluted with deionized water at a flow rate of 0.2 ml/min.
9. The polymer–protein conjugate is eluted in 0.3 M sodium phosphate buffer pH 6.8.

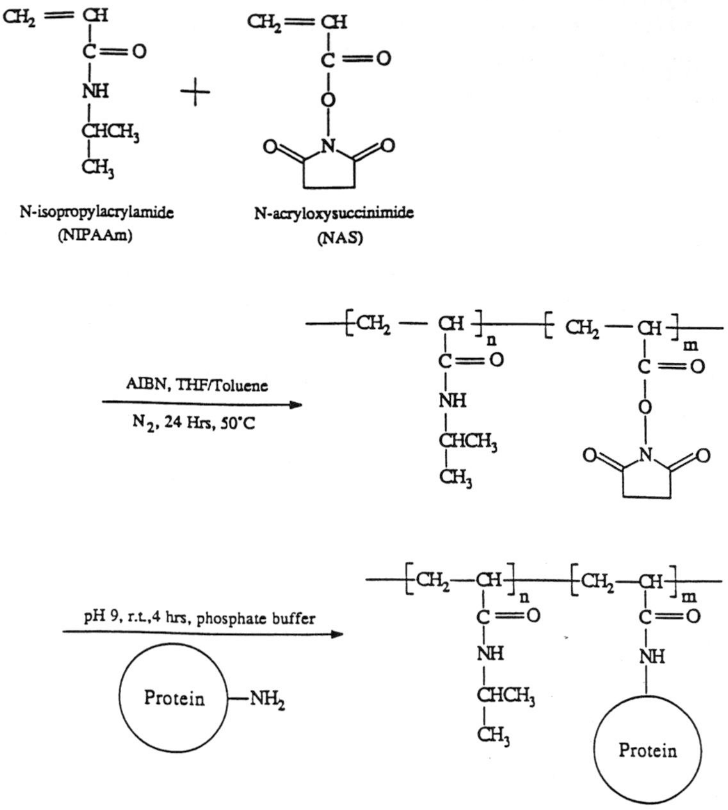

Figure 5. The synthesis of a protein conjugate with an *N*-isopropylacrylamide/*N*-acryloxysuccinimide copolymer.

4.2 Site-specific conjugation to engineered proteins

Protocol 5. Maleimide conjugation reaction

Reagents

- 50 mM Sodium phosphate, 1 mM EDTA ph 8
- Dithiothreitol
- Sephadex G-25

- Maleimide-terminated oligo(NIPAAm) prepared as in *Protocol 2*
- Saturated ammonium sulfate

Method

1. 160 μl of a 1 mM protein (containing a free cysteine side chain) solution in 50 mM phosphate, 1 mM EDTA buffer pH 8 is reduced with 1 mM dithiothreitol (DTT) for 10 min at 4°C.

2. The mixture is passed over a Sephadex G25 gel filtration column equilibrated with the same buffer to recover the protein free of DTT. The protein band is collected in a 15 ml centrifuge tube containing a tenfold molar excess of MI-oligo(NIPAAm).

3. The reaction is allowed to proceed for 4 h at room temperature with gentle shaking to ensure complete mixing of the reactants.

4. The protein/MI-oligo(NIPAAm) conjugate was separated from unreacted protein by adding 10% (v/v) saturated $(NH_4)_2SO_4$ to the mixture to depress the lower critical solution temperature (LCST), i.e. cloud-point, from 32°C to ∼ 20°C, and the reaction mixture is warmed to 30°C (> LCST) to selectively precipitate the protein/MI-oligo(NIPAAm) conjugate.

5. The precipitate is then separated by centrifugation (10 000 *g*, ambient temperature) to produce a pellet (pink-coloured in the case of the haem protein). The pellet is redissolved in buffer and the precipitation step was repeated twice to ensure complete removal of the unreacted protein.

Protocol 6. Vinyl sulfone conjugation reaction

Reagents

- PBS
- Vinyl sulfone-poly(NIPAAm) prepared as in Protocol 3
- Tris(2-carboxyethyl) phosphine (TCP)

Method

1. The conjugation of the vinyl sulfone–poly(NIPAAm) to the streptavidin/N49C mutant was performed at pH 7 in PBS, with Tris(2-carboxyethyl) phosphine (TCEP) as a disulfide reducing agent. A large excess of polymer (molar ratio of polymer:protein of 50:1) is used. A second aliquot of the TCEP is added after 30 min.

2. The reaction is incubated at 25°C for 4 h and then allowed to react at 4°C for an additional 20 h.

3. The polymer–protein conjugate is separated from unreacted protein by thermal precipitation of the poly(NIPAAm) at 37°C, which phase separates the streptavidin conjugate.

Protocol 6. *Continued*

4. Characterization by HPLC analysis using a gel permeation column and polyacrylamide gel electrophoresis (SDS–PAGE) is used to demonstrate that all of the streptavidin purified in this manner is conjugated to poly(NIPAAm).

4.3 Characterization of conjugates by mass spectrometry

The conjugates can be characterized by mass spectrometry. As an example, we have characterized the T8C cytochrome $b5$ conjugates by MALDI-TOF mass spectrometry. The spectra are acquired on a Finnigan MAT Laser-MAT™ LD-TOF instrument. Samples are prepared by depositing ~ 1 ml of sample on the centre of a gold plated metal target. The T8C/MI-oligo (NIPAAm) conjugate is at a concentration of 1 mM. The matrix used was 2,5 dihydroxybenzoic acid, dissolved in 60% acetonitrile, 40% 0.1% TFA at a concentration of ~ 5–10 mg/ml. The molar ratio of analyte to matrix is typically 1:10 000. The samples are allowed to air dry and placed in the spectrometer. Multiple laser shot spectra are acquired after spectral optimization, which involves determining the minimum laser power density required to observe analyte ions from one of four predefined target positions. The instrument is calibrated using sperm whale apomyoglobin as a calibrant.

The T8C/MI-oligo NIPAAm conjugate shows peaks at ~ 11 000 Da [(M + H)$^+$], and a minor peak at ~ 13 000 Da. The latter peak is not observed in the spectrum of protein alone (native $b5$, T8C) or the spectrum of a physical mixture of native $b5$ and MI-oligo(NIPAAm). The MALDI-TOF MS spectrum of MI-oligo(NIPAAm) displays a range of peaks differing by 113 Da, suggesting that they can be assigned to [nM + H]$^+$ ions where M is the monomer (minus the maleimide end-group). The observation of these peaks is consistent with the distribution of oligomer chain lengths expected from free radical polymerization. More importantly, however, the molecular weight distribution of the oligomer is centred at ~ 1900 Da, which is consistent with the difference in mass of the protonated protein and the unique peak at ~ 13 000 Da (11 000 + 1900 Da) observed only in the spectrum of the conjugate. These results strongly support an oligomer:protein stoichiometry of 1:1.

References

1. Hoffman, A. S. (1991). *MRS Mater. Res. Soc. Bull. XVI*, **9**, 42.
2. Hoffman, A. S. (1995). *Macromol. Symp.*,
3. Monji, N. and Hoffman, A. S. (1987). *Appl. Biochem. Biotechnol.*, **14**, 107.
4. Chen, J. P. and Hoffman, A. S. (1990). *Biomaterials*, **11**, 631.
5. Chen, G. H. and Hoffman, A. S. (1993). *Bioconjugate Chem.*, **4**, 509.
6. Chen, G. H. and Hoffman, A. S. (1994). *J. Biomater. Sci. Polym. Ed.*, **5**, 371.
7. Park, T. G. and Hoffman, A. S. (1993). *Enz. Microb. Technol.*, **15**, 476.

8. Miura, M., Cole, C. A., Monji, N., and Hoffman, A. S. (1991). *The 17th Annual Meeting of the Society for Biomaterials*. Scottsdale, Arizona.
9. Wu, X. S., Hoffman, A. S., and Yager, P. (1992). *Polymer*, **33**, 4659.
10. Chilkoti, A., Chen, G., Stayton, P. S., and Hoffman, A. S. (1993). *Bioconjugate Chem.*, **5**, 504.
11. Stayton, P. S., Shimoboji, T., Long, C., Chilkoti, A., Chen, G., Harris, J. M., *et al.* (1995). *Nature*, **378**, 472.
12. Dong, L. C. and Hoffman, A. S. (1991). *J. Control Release*, **15**, 141.
13. Brannon-Peppas, L. and Peppas, N. A. (1989). *J. Controlled Release*, **8**, 267.

9

Kinetic analysis of the interaction between an analyte in solution and an immobilized protein

R. KARLSSON and S. LÖFÅS

1. Introduction

Proteins function by interacting with other biomolecules. On a molecular basis the interaction can be characterized in terms of the affinity and the kinetics of the interaction. The affinity is often easy to measure (1) and describes the strength of the interaction. Kinetic analysis is more informative than affinity analysis and gives additional data regarding the rate of complex formation and dissociation. When kinetic analysis is possible, the binding assay is facilitated and fewer experiments can be performed. This is illustrated in *Figure 1*. Here data obtained using a traditional solid phase assay (i.e. an ELISA) and a biosensor-based kinetic assay is compared. When a single concentration of analyte is used a traditional assay gives one data point that in itself is almost meaningless. In contrast a biosensor assay can provide direct and continuous information on how the analyte binds to and dissociates from the immobilized ligand. In theory a single binding curve can be used to determine both the association rate constant, k_a, and the dissociation rate constant, k_d. The affinity constant K_A can then be calculated from the ratio between k_a and k_d.

Kinetic information is vital when the effect of a change in the primary structure of a protein on its function is evaluated (2–5) or when different reagents have to be selected for assay purposes. Kinetic analysis is often performed with a combination of stopped flow techniques and different forms of spectroscopy (6, 7) and with both binding partners present in solution. Labelling and careful calibration of the detection system is often required. In contrast optical sensors (8, 9) permit direct detection of binding events taking place at a sensor surface. No labelling is required for detection. Kinetic analysis of interactions where one biomolecule is attached to a sensor surface and the other binding partner is present in solution therefore becomes possible (10). The usefulness of optical sensors for kinetic analysis is determined by a number of parameters related to the detection system, the sensor surface, and on how the analyte molecule is presented to the surface.

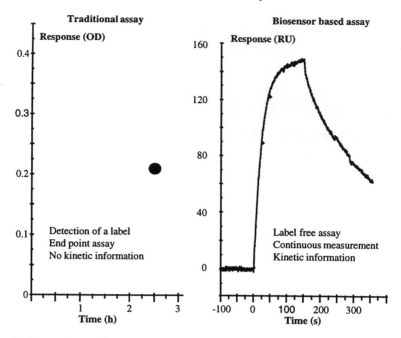

Figure 1. Comparison of data obtained using a traditional solid phase assay and a biosensor-based assay. In both instances one concentration of analyte reacts with the surface immobilized ligand.

2. Kinetic analysis on immobilized proteins

One obvious requirement for kinetic analysis on immobilized proteins to work is that the immobilization procedure leaves at least a fraction of the immobilized molecules in active form.

2.1 The ideal case

In an ideal case, all immobilized molecules are equally active and all binding sites are independent. When [B] is the concentration of the immobilized ligand, [A] is the concentration of the binding partner (the analyte) in solution, and [AB] is the concentration of the complex formed when A reacts with B, then:

$$d[AB]/dt = k_a \times [A] \times [B] - k_d \times [AB] \qquad [1]$$

where k_a is the association rate constant and k_d is the dissociation rate constant. Thus, to obtain kinetic information from a surface reaction it must be possible to:

(a) Determine the change in the concentration of AB complex.

(b) Know the concentration of the analyte A.

The concentration of B at any given time is equal to the total concentration of B minus the concentration of AB complex. The analyte concentration can often be regarded as constant and equal to C_A during the experiment. Therefore:

$$d[AB]/dt = k_a \times C_A \times [B_{tot} - AB] - k_d \times [AB] \qquad [2]$$

Integrating *Equation 2* gives:

$$[AB(t)] = [AB_{eq}] - [AB_{eq}] \times e^{-(k_a \times C_A + k_d) \times t} \qquad [3]$$

where $[Ab_{eq}]$ is the equilibrium concentration of AB complex. At equilibrium $d[AB]/dt$ equals zero and consequently:

$$[AB_{eq}] = \frac{k_a \times C_A \times B_{tot}}{k_a \times C_A + k_d} \qquad [4]$$

When the interaction is interrupted and analyte replaced with buffer, the AB complex starts to dissociate. When the concentration of released A can be neglected (i.e. $C_A = 0$) the dissociation event is described by:

$$[AB(t)] = [AB(t_0)] \times e^{-k_d \times t} \qquad [5]$$

where $AB(t_0)$ is the concentration of AB complex at time t_0 when analyte is replaced with buffer.

An important consequence of *Equations 3* and *5* is that plots of $\ln(d[AB]/dt)$ versus t are linear. Typical progress curves for the ideal case, the linear transforms, and the terminology are demonstrated in *Figure 2*.

2.2 More complex interactions

In practice ideal binding curves are not always obtained. Deviations from the idealized case are often easy to detect since $\ln(d[AB]/dt)$ versus t plots are no longer linear. The cause of these deviations can vary from case to case but can often be explained quite rationally. For instance:

(a) The analyte or another compound present in the analyte solution can bind non-specifically to either B or to the surface itself.

(b) The analyte is heterogeneous and exists in multiple forms with different binding properties.

(c) The analyte has multiple binding sites and multivalent binding becomes possible.

(d) The concentration of analyte changes due to sample depletion and cannot be considered constant over time.

(e) The binding rate is so fast that the analyte concentration close to the surface cannot be maintained during the reaction.

(f) The immobilized ligand B is heterogeneous and exists in multiple forms with different binding properties.

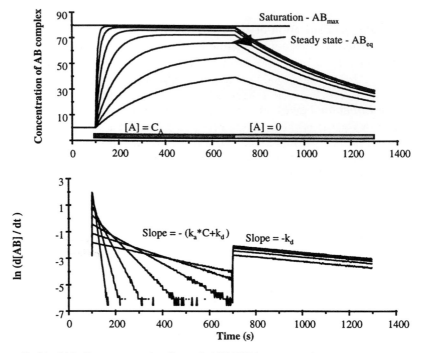

Figure 2. Ideal binding curves give linear ln(d[AB]/d*t*) versus *t* plots.

(g) Immobilization through multiple sites has rendered B heterogeneous.

(h) The interaction proceeds through intermediate steps until a stable complex is formed.

Many of these points can be clarified by doing control experiments (a), by checking the homogeneity of A and B with other techniques (b, f), by varying the concentrations of A (f) or B (e), or by varying the interaction time (b, h), or by testing different immobilization chemistries (g).

3. Monitoring binding events on surfaces

When an analyte molecule binds to a surface immobilized ligand, the mass on the surface will increase, and as a consequence, the refractive index of the layer where binding takes place will also increase. Techniques that measure changes in refractive index at the surface are therefore potentially useful for monitoring binding events at surfaces (11). One such technique based on surface plasmon resonance detection combined with modified gold surfaces and a flow system for sample delivery will be described. This technique is commercially available in BIACORE instrumentation.

3.1 Surface plasmon resonance (SPR)

Surface plasmon resonance is an optical phenomenon. Light incident at different angles through a prism hits a thin (50 nm) metal film, usually gold or silver (*Figure 3*). The other side of the metal film is in contact with the buffer solution. The intensity of the reflected light is monitored as a function of the incident angle. When surface plasmon resonance occurs, light energy is lost to the metal film and the intensity of the reflected light drops. The resonance phenomenon occurs for light incident at a sharply defined angle. The position of the minimum of reflected light intensity depends on changes in refractive index near the interface on the buffer side. Binding of an analyte to the gold film will lead to a change in refractive index, and a plot of the position of the light intensity minimum versus time directly reflects the binding event. A change in buffer composition can also give rise to a change in refractive index. Fortunately, the optical system can be configured so that several spots on a single surface can be monitored simultaneously. By using a spot with no ligand immobilized, the buffer effect can therefore be monitored separately and corrected for.

For a binding event, the change in refractive index is proportional to the change in mass (12). The SPR signal, R, is therefore proportional to the concentration of complex formed and can replace AB in *Equations 1–5*.

Similarly the maximum SPR response R_{max}, corresponding to the situation where all binding sites are saturated, is proportional to the total concentration, B_{tot}, of the immobilized partner. R_{max} can therefore be used instead of

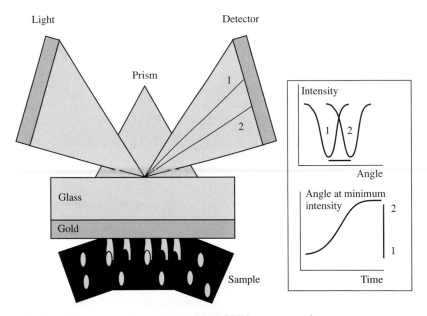

Figure 3. The SPR configuration used in BIACORE instrumentation.

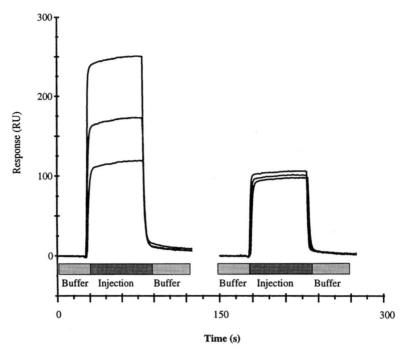

Figure 4. When the analyte–ligand complex dissociates rapidly binding curves resemble square waves. When analyte is injected over surfaces with varying levels of immobilized ligand a true binding event (left panel) is easy to recognize since the response level is proportional to the level of immobilized ligand.

B_{tot} in *Equations 2* and *4*. Typically one to ten data points are recorded every second and detailed binding curves are obtained. In *Figures 4* and later the response is given in resonance units, RU. This unit is used in BIACORE instrumentation, and one RU approximately corresponds to a mass increase of 1 pg/mm^2 surface.

For studies involving low molecular weight analytes and for low affinity interactions it is a clear advantage to monitor simultaneously binding events on spots with different levels of immobilized ligand. In the first case small signals can be expected and in the second case dissociation is often very rapid (*Figure 4*). In both instances, a true binding event is easier to discriminate from background signals on a multispot surface since the amplitude of the response is proportional to the immobilization level (13).

4. Establishing contact between sample and surface

4.1 Kinetics and the use of a flow system

Kinetic analysis is simplified when the concentration of analyte is constant during the interaction. It is also an advantage to be able to switch rapidly from

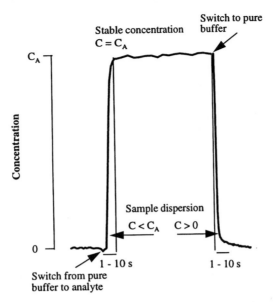

Figure 5. The switch from pure buffer to analyte solution and back to pure buffer should be rapid so that dispersion effects occur over short time frames. This will reduce concentration errors.

having pure buffer in contact with the surface to analyte conditions and back to pure buffer conditions again. The analyte concentration is uncertain over the time frame when liquids are exchanged and this period should be as short as possible. Since refractive index varies with temperature it is also important to control the temperature in order to minimize drift.

When the sample is introduced in a flow system, a complete switch from pure buffer to analyte solution only takes a few seconds (*Figure 5*). The analyte solution is continuously replenished. This means that sample depletion is reduced to such an extent that the concentration of analyte can be considered constant and equal to C_A during injection. When pure buffer again flows over the surface the analyte cannot accumulate there as the AB complex dissociates and C_A is equal to zero. Small dimensions allow for a rapid temperature equilibration, and when the sample can be injected serially over several measuring spots, a reference surface can be used to compensate for any remaining temperature drift.

4.2 The balance between reaction rate and diffusion rate

When the concentration of analyte is constant in the entire system (*Figure 6*), true kinetics are measured. This means that *Equations 1–5* are valid for an interaction involving single and independent sites on both analyte and ligand.

For very fast reactions, the concentration criterion is not always valid. Instead the concentration of analyte close to the surface may drop and fall

The kinetic situation

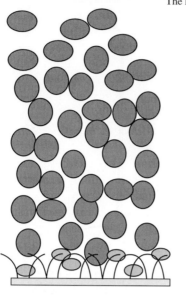

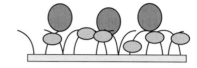

During injection $C_s = C_A$ After injection $C_s = C_A = 0$

C_A = concentration of injected analyte
C_s = concentration of free analyte at the surface

Figure 6. In a true kinetic situation, the concentration of analyte is the same in the entire system. The concentration is C_A when analyte is present and zero when buffer has replaced the analyte solution.

below the concentration in the bulk solution. This happens when the binding rate is larger than the rate of analyte diffusion to the surface. As a consequence, the observed binding rate becomes lower than the intrinsic binding rate. Similarly, the observed dissociation rate in the buffer flow phase is also reduced compared to the intrinsic dissociation rate. This is because analyte molecules can rebind before they are washed away from the surface (*Figure 7*).

At higher flow rates, the thickness of the depleted layer is reduced. The diffusion distance becomes shorter, the concentration close to the surface increases, and this leads to more rapid binding. Deviations from true kinetics due to limited diffusion are therefore very easy to detect by varying the flow rate. By lowering the concentration of the immobilized ligand, the balance between the intrinsic binding rate and the diffusion limited binding rate is improved. Still this balance governs the range of association rate constants that can be accurately determined without the need for correction factors. The expected range in the BIACORE instrumentation at a maximum binding capacity of 100 RU and a flow rate of 30 μl/min is illustrated in *Figure 8*. This figure is based on a comparison of binding rates obtained for interactions

Kinetics and mass transfer

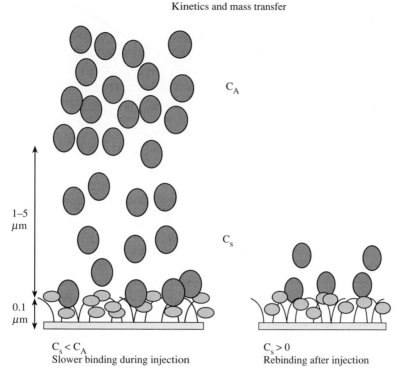

C_A

C_s

1–5
μm

0.1
μm

$C_s < C_A$
Slower binding during injection

$C_s > 0$
Rebinding after injection

Figure 7. In a diffusion limited situation the concentration of analyte is not identical in the entire system.

that are truly kinetic with binding rates calculated under diffusion limited conditions (14–16).

When k_a values fall below the lower curve, near ideal kinetic data can be obtained. In the mid region binding rates are partly limited by diffusion. Diffusion limitations can be incorporated in the analysis using numerical integration (17) and accurate k_a values in this region can then be determined. This and the observation that dissociation rate constants in the range from 10^{-1} sec^{-1} to 10^{-5} sec^{-1} can be determined suggests that kinetic analysis of most protein–protein interactions will be possible (18, 19). Only very high k_a values will be difficult to determine, particularly for low molecular weight analytes. This is because detection is mass-sensitive, and high concentrations (in moles/mm^2) of immobilized ligand must be used to detect binding of a low molecular weight analyte. When the experiment is redesigned so that a high molecular weight and low molecular weight analyte compete for the binding site of the immobilized receptor, the density of the immobilized ligand can be kept low, and a wider range of rate constants for low molecular weight analytes appears to be possible (20, 21).

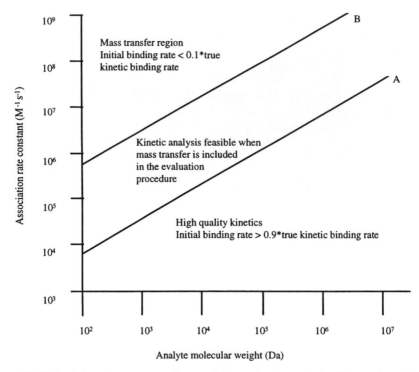

Figure 8. Plot of k_a values versus the analyte molecular weight. For each molecular weight, the figure describes the balance between the kinetic situation and the diffusion limited situation.

5. Surface properties

In kinetic studies of interactions between immobilized proteins and analyte in solution, the possible influences from the sensor surface must be considered. The surface type as well as the immobilization procedure will determine the activity of the immobilized ligand and this in turn can influence the binding kinetics. Sensor surface properties to be considered include the surface structure, the degree of hydrophobicity or hydrophilicity, the surface charge, and the type of functional groups used for immobilization. The amounts of immobilized ligand as well as type of immobilization procedure and number of attachment points to the sensor may also have impact on the activity of the ligand. For instance, most proteins will adsorb in a random manner on flat, rigid surfaces leading to different states of binding activities due to steric hindrance or blockage. Depending on the type of surface, non-covalent surface binding can also induce denaturation of the protein (22). Even when functional groups for site-directed covalent coupling are introduced on the surface, the intrinsic physical properties of the surface will to a large extent govern the orientation and activity of the immobilized protein (23).

The use of SPR technology for biomolecular interaction studies may at first appear limited since SPR relies on surfaces with metal films (gold or silver). A wide range of surface modifications have been tested and found compatible with the SPR technology. However, the most reliable and well characterized way of gold film modification is based on the formation of self-assembled monolayers (SAM) on the metal surface. In particular, alkanethiols are used (24). Typically, dilute solutions of the alkanethiol are contacted with the metal and by chemisorption via metal–sulfur bond formation, a monomolecular dense film is spontaneously formed. By using various tail groups in the alkanethiol, surfaces with different properties can be created (25). These groups can be used for direct immobilization or for introduction of other functional groups. Modification with dextran polymer chains in order to convert the flat surface to a three-dimensional and flexible structure has proved to be successful for studies of biomolecular interactions with SPR (26). This highly hydrophilic hydrogel-like layer has low non-specific binding, and immobilization by different methods (27, 28) gives ligands with highly preserved binding activities (29). A further advantage is that immobilization to the flexible dextran polymer instead of directly to the solid surface largely eliminates the need of oriented coupling. Surfaces adapted to support lipid layers with embedded ligands have recently been described (30, 31).

6. Immobilization procedures

The amines, carboxylic acids, and thiols in corresponding amino acid residues are most commonly used for protein immobilization (32). In addition, proteins can be attached in a non-covalent manner via affinity capture to a covalently immobilized reagent. This reagent can be an antibody directed towards a part of the protein separate from the interaction site. Immobilization via avidin or streptavidin represents a hybrid of these two methods in that it normally requires a covalent modification of the protein by a biotinylation reagent.

The carboxylated dextran sensor surface provided with BIACORE instrumentation is primarily intended for direct covalent immobilization via amine groups. A fraction of the carboxylic groups are converted to reactive esters via *in situ* reaction with a mixture of a water soluble carbodiimide (EDC) and N-hydroxysuccinimide (NHS). By introducing the protein in a low ionic strength buffer with a pH below the isoelectric point of the protein, it will concentrate within the dextran layer through interaction with residual negative charges. At the same time it will react with newly formed ester groups (33). Thus, very efficient immobilization can be achieved at protein concentrations as low as 5–10 µg/ml. The immobilization procedure can be monitored directly with SPR (*Figure 9*) and the amount of bound protein is easily determined.

The carboxylic groups in the dextran layer can be converted to thiols or

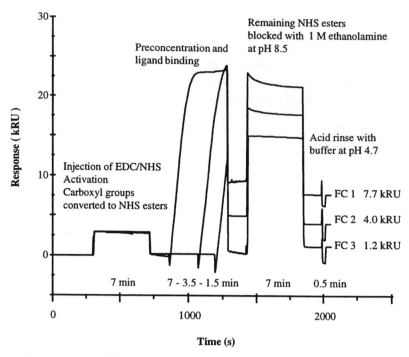

Figure 9. Covalent immobilization of protein (IgG) to a surface with carboxyl groups followed in BIACORE instrumentation. The carboxyl groups are converted to active esters and react with amine groups on the protein. By varying the protein injection time different levels of protein are immobilized in different flow cells.

thiol-reactive groups for selective coupling, or to hydrazide groups for use in condensation reactions involving ligands with aldehyde functions (28, 33). These coupling methods provide alternative paths to site-directed immobilization and are valuable in situations where the ligand lacks reactive amines, like in small peptides, or when active amines are present in the active site and important for the interaction with the analyte.

The activity of the immobilized ligand can be estimated by comparing the surface amount of the ligand with the saturation response obtained from interaction with the analyte (*Figure 10*). The activity of the immobilized ligand is routinely high (> 70%) (29) even when employing the random coupling via amine groups in lysine residues. Introduction of reagents by capturing techniques can often give higher activities (34).

7. The kinetic experiment

When the ligand has been immobilized, a few preliminary experiments are useful before the kinetic parameters of the interaction are determined.

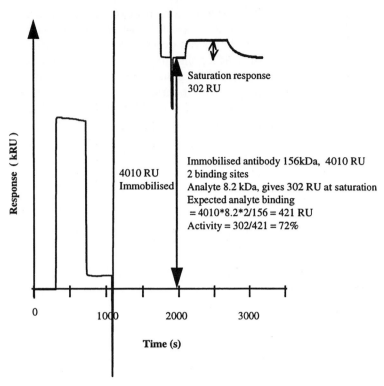

Saturation response
302 RU

Immobilised antibody 156kDa, 4010 RU
2 binding sites
Analyte 8.2 kDa, gives 302 RU at saturation
Expected analyte binding
= 4010*8.2*2/156 = 421 RU
Activity = 302/421 = 72%

4010 RU
Immobilised

Response (kRU)

Time (s)

Figure 10. The SPR response is proportional to mass. The activity of the immobilized ligand can be calculated from the immobilization level and the saturation response.

The specificity of the interaction can be checked by looking at non-specific binding to a control surface lacking the immobilized ligand and by performing inhibition experiments where the analyte is pre-mixed with the ligand in solution. With high concentrations of ligand in solution, it should be possible to completely block the binding of analyte to the immobilized ligand.

It has been demonstrated that diffusion limitations can pass unnoticed and lead to over-interpretation of the data (15, 35). This risk is easily eliminated by varying the flow rate. When the binding rate is limited by diffusion, it will depend on the flow rate. In a severe case of diffusion limitation, the initial binding rate will increase almost three times when the flow rate is raised from 3 to 75 µl/min. Furthermore, the outcome of an experiment where the flow rate is varied will guide the user to proper data analysis. In the absence of a flow rate effect (< 10% variation in binding rate for flow rates 15 and 75 µl/min), data can be analysed using 'true kinetic' models. This is an advantage since fewer parameters have to be calculated. Usually, the first assumption will be that *Equation 1* can be used in data analysis. When diffusion limitations are observed, it may be a good idea to repeat the experiment with less ligand immobilized. When this is not possible or when the diffusion limitation cannot be overcome, an equation system describing both the diffusion to the surface and the reaction at the surface should be used in data analysis (36–38).

Detailed kinetic analysis involves injection of several concentrations of analyte. If the dissociation of analyte from the analyte–ligand complex is slow ($k_d < 1 \times 10^{-3}$ sec^{-1}), it will be impractical to wait for complete dissociation and it will be necessary to regenerate the surface. The regeneration procedure should speed up dissociation without affecting the activity of the ligand. Many protein–protein interactions are sensitive to pH or to the presence of chaotropic salts and regeneration conditions are often found in short time.

For kinetic analysis in BIACORE the following guidelines are useful:

(a) Use a surface with a level of immobilized ligand to obtain a favourable balance between the reaction rate and the diffusion rate. A saturation response of 50–200 RU is appropriate.

(b) Inject a low concentration of analyte and determine the initial binding rate at flow rates of 3, 15, and 75 μl/min. Use a low concentration of the analyte so that binding rates far from steady state conditions can be measured.

(c) Inject analyte for 1–10 min and monitor dissociation of the analyte ligand complex for 5–30 min.

(d) Use analyte concentrations ranging from $0.5 \times K_D$ to $100 \times K_D$. The K_D concentration is the concentration that gives an equilibrium response equal to half the saturation response.

These are practical guidelines, and further experiments involving surfaces with different levels of immobilized ligand (39) or varying injection time (37) can be required, in particular when deviations from the ideal case as described in Section 2.2 are encountered.

8. Data analysis

8.1 Qualitative aspects

Prior to computer aided data analysis, it is useful to inspect the binding curves in an overlay plot. This ensures a rapid overview of binding data. In *Figure 11* it is easy to see that binding is concentration-dependent, that steady state is reached, that dissociation is rapid, and that there are no obvious outliers.

The overlay plot is also useful for rapid comparisons. Two examples are shown in *Figure 12*. In the left panel the binding curves reveal that the point mutations introduced lead to variants that dissociate more rapidly than the wild-type protein. The right panel illustrates the effect of temperature on binding kinetics. In this case the affinity is lower at higher temperatures and dissociation is more rapid.

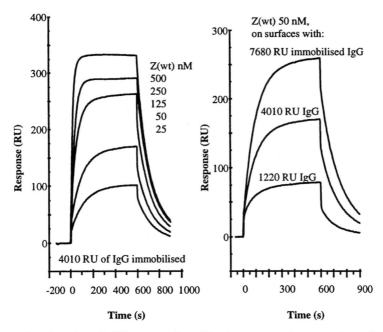

Figure 11. Overlay plot of different analyte (Z(wt)) concentrations reacting with immobilized ligand (IgG) (left) and overlay plot of one concentration of analyte (Z(wt)) reacting with ligand (IgG) immobilized to three different levels (right). Z(wt) is a one domain analogue of protein A and binds to the Fc part of IgG.

8.1 Quantitative aspects

For quantitative analysis of binding data the reaction mechanism and an appropriate interaction model must first be considered. The models describe the reaction with a number of parameters (a1,a2,a3..) such as time, the rate constants, the concentration of immobilized ligand, the concentration of analyte, and the concentration of the analyte–ligand complex. In non-linear regression analysis (40) values for the unknown parameters (typically the rate constants) are varied. Binding curves are calculated from the model and compared with the observed binding curve. When the difference between measured and calculated response curves (the residual) is minimized, the best fit curve is obtained and calculated parameter values are returned. This iterative procedure is illustrated in the left panel of *Figure 13*. When observed and calculated data are close the model is sufficient to describe the interaction data but this does not prove that the model is correct. Calculated rate constants are therefore apparent rate constants and should be reported as such.

Non-linear analysis of kinetic data can be used to various degrees of sophistication. The injection phase and the buffer flow phase can be analysed separately (41) using the appropriate analytical equations. Simultaneous analysis, or global fitting of all data, not only from one binding curve but including all

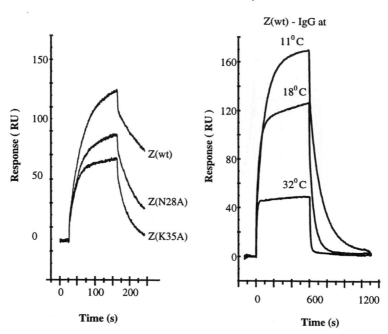

Figure 12. Overlay plot illustrating how one concentration of different analytes—Z wild-type and mutants—react with the same immobilized ligand (left panel) and overlay plot illustrating the effect of temperature on the binding kinetics of the wild-type (right panel).

the curves in the experiment is more attractive (42, 43). In global analysis the residual is minimized for the entire data set and not only for one curve at a time. This gives more reliable results, and a practical aspect is that only one set of rate constants is calculated instead of one set for each curve. The right panel in *Figure 13* demonstrates the result from a global analysis of the data in *Figure 11*. The binding curves were analysed using the one to one model described by *Equation 1*. The plot is an overlay plot of measured and calculated data. Clearly the one to one model describes this data set adequately.

When complex interaction models are used analytical solutions to the differential rate equations are often difficult if not impossible to find. By using numerical integration the response curves from complex models can easily be calculated and later analysed. For kinetic analysis a combination of numerical integration of differential rate equations with non-linear analysis and global fitting appears to be very useful (44).

9. Kinetic analysis with BIACORE

The kinetic experiments are straightforward. The ligand is immobilized and the analyte is injected. In addition a flow rate experiment is useful to identify a situation where the binding rate is limited by diffusion to the surface.

164

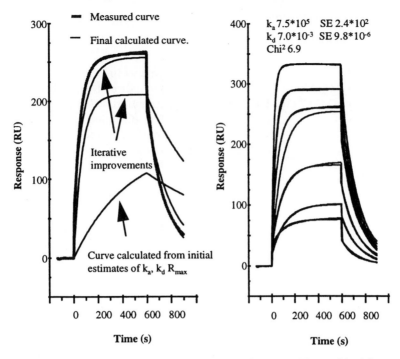

Figure 13. The iterative process of data analysis (left panel). The residual is gradually minimized and the two upper curves are almost identical. These curves are the measured curve and the final result from data analysis. The right panel is an overlay plot of measured and calculated data. The measured data is the same as in *Figure 11*.

Protocol 1. Immobilization

Equipment and reagents

- BIACORE 2000 instrumentation: this instrumentation includes automatic buffer and sample delivery by pumps and an autoinjector. Samples are placed in vials and put in sample racks. Each sample position can be addressed by the autoinjector. Samples can be injected in four flow cells and each flow cell contacts the sample with approx. 1 mm² of the sensor surface. The instrument is controlled from a PC. Injection sequences can be programmed and executed automatically.

- Sensor surface CM5: this is a sensor surface with carboxylated dextran attached to the gold film
- HBS buffer: 10 mM Hepes, 0.15 M NaCl, 0.005% Tween 20 pH 7.4
- 0.2 M EDC, 50 mM NHS, and 1 M ethanolamine–HCl pH 8.5
- Ligand at 10–100 μg/ml in 10 mM buffer at a pH 0.5–1.5 unit lower than pI

Method

1. Insert the sensor surface and prime the system by pumping HBS through the flow system and over the sensor surface.

Protocol 1. *Continued*

2. Use the following program for immobilization.

```
DEFINE APROG immob
FLOW 10                  ! Sets the flow rate to 10 μl/min.
FLOWPATH 1,2,3           ! Directs the flow to three flow cells (spots).
DILUTE r1e1 r1e2 r1e3 50 ! EDC is in rack position r1e1 and NHS in rack
                           position r1e2. The autoinjector will transfer
                           equal volumes of EDC and NHS to an empty
                           tube in rack position r1e3 and mix them.
INJECT r1e3 70           ! Injects 70 μl of the EDC/NHS mixture.
INJECT r1e4 70,35,15     ! Injects ligand from position r1e4. 70 μl is
                           injected in flow cell 1, 35 μl in flow cell 2, and
                           15 μl in flow cell 3.
INJECT r1e5 70           ! Injects 70 μl ethanolamine from position r1e5.
INJECT r1f3 5            ! Injects 5 μl of the regeneration solution (see
                           below).
END
MAIN                     ! Will run listed Aprogs.
FLOWCELL 1,2,3,4         ! Sets data collection to all four flow cells.
APROG immob              ! Selected aprog.
END
```

3. See *Figure 9* for experimental data.

Protocol 2. Kinetic analysis

Equipment and reagents

- Analyte diluted in running buffer, placed in rack positions r1a1 to r1a5
- Regeneration solution placed in position r1f3

Method

1. Use the following program for kinetic analysis.

```
DEFINE APROG Flow        ! Test for a diffusion limitation.
PARAM %flow %pos1 %vol   ! Parameter names. Values obtained from
                           loop table.
FLOW %flow               ! Sets the flow defined in the flow loop table.
FLOWPATH 3               ! Samples will be injected in flow cell 3 where
                           the least amount of ligand is immobilized.
INJECT %pos1 %vol        ! An injection with low dispersion. Inject
                           positions and volumes are found in the
                           flow loop table.
FLOW 15
INJECT r1f3 15
END
```

166

```
DEFINE LOOP Flow
LPARAM %flow %pos1 %vol
TIMES 1                          ! The loop will be used once. TIMES 2 will
                                   give duplicates.
        3       r1a1    6        ! With these flow rates and injection
        15      r1a1    30       volumes the injection time is 2 min.
        75      r1a1    150
END
DEFINE APROG Kinetics
PARAM %pos1 %conc
Keyword conc %conc               ! Keywords are stored with the data and
                                   a help to identify important conditions.
                                   In this case the concentration of the
                                   analyte.

FLOW 20
FLOWPATH 1,2,3
WAIT 300                         ! Allows the baseline to stabilize for 300 sec.
KINJECT %pos1 200 300            ! Injects 200 µl from a position defined in
                                   the loop table and allows for a 300 sec
                                   buffer flow phase before wash of the
                                   injection system.

INJECT r1f3 10
END
DEFINE LOOP Kinetics
LPARAM %pos1 %conc
TIMES 1
            r1a1    25_nM
            r1a2    50_nM
            r1a3    125_nM
            r1a4    250_nM
            r1a5    500_nM
END
MAIN                             !This program will run the flow test and
FLOWCELL 1,2,3,4                 the concentration series defined in the
LOOP Flow ORDER                  Flow and Kinetics programs.
APROG Flow %flow %pos            ! Flow rates will be used in the
   1 %vol                        programmed ORDER.
ENDLOOP                          ! Whereas analyte concentrations will be
LOOP Kinetics RANDOM             injected in RANDOM order.
APROG Kinetics %pos1 %conc
ENDLOOP
END
```

2. See *Figure 11* for experimental data obtained with Aprog Kinetics.

10. Concluding remarks

Direct and label-free sensing of binding events on immobilized surfaces provides a wealth of information. Data obtained can be used to characterize and optimize surface properties such as protein binding capacity and the degree of non-specific binding. When analytes react with the immobilized ligand the affinity and kinetic properties of the interaction can be determined. Binding of modified analytes can easily be compared directly using overlay plots of binding curves. The range of rate constants that can be determined is limited by the balance between the reaction rate and the rate at which analyte diffuse to the surface. Improvements in instrument sensitivity and in data evaluation methods now allows kinetic analysis of most protein–protein interactions. With appropriate experimental design it even seems possible to simultaneously determine both the active concentration of the analyte and the kinetic properties of the analyte ligand interaction (34, 45, 46).

References

1. Phizicky, E.M. and Fields, S. (1995). *Microbiol. Rev.*, **59**, 94.
2. Cunningham, B.C. and Wells, J.A. (1993). *J. Mol. Biol.*, **234**, 554.
3. van der Merwe, P.A., McNamee, P.N., Davies, E.A., Barclay, A.N., and Davis, S.J. (1995). *Curr. Biol.*, **5**, 74.
4. Jendeberg, L., Persson, B., Andersson, R., Karlsson, R., Uhlén, M., and Nilsson, B. (1995). *J. Mol. Recog.*, **8**, 270.
5. Barbas III, C.F., Hu, D., Dunlop, N., Sawyer, L., Cababa, D., Hendry, R.M., *et al.* (1994). *Proc. Natl. Acad. Sci. USA*, **91**, 3809.
6. Gaikwad, A., Gómez-Hens, A., and Pérez-Bendito, D. (1993). *Anal. Chim. Acta*, **280**, 129.
7. Goldberg, M.E. (1991). *Trends Biol. Sci.*, **16**, 358.
8. Jönsson, U. and Malmqvist, M. (1992). *Adv. Biosens.*, **2**, 291.
9. Cush, R., Cronin, J.M., Stewart, W.J., Maule, C.H., Molloy, J., and Goddard, N.J. (1993). *Biosens. Bioelectron.*, **8**, 347.
10. Karlsson, R., Michaelsson, A., and Mattson, L. (1991). *J. Immunol. Methods*, **145**, 229.
11. Brecht, A. and Gauglitz, G. (1995). *Biosens. Bioelectron.*, **10**, 923.
12. Stenberg, E., Persson, P., Roos, H., and Urbaniczky, C. (1991). *J. Colloid Interface Sci.*, **143**, 513.
13. Karlsson, R. and Ståhlberg, R. (1995). *Anal. Biochem.*, **228**, 274.
14. Sjölander, S. and Urbaniczky, C. (1991). *Anal. Chem.*, **63**, 2338.
15. Glaser, R.W. (1993). *Anal. Biochem.*, **213**, 152.
16. Karlsson, R., Roos, H., Fägerstam, L., and Persson, B. (1994). *Methods: a companion to methods in enzymology*, **6**, 99.
17. Morton, T.A., Myszka, D.G., and Chaiken, I.M. (1995). *Anal. Biochem.*, **227**, 176.
18. Northrup, S.H. and Erickson, H.P. (1992). *Proc. Natl. Acad. Sci. USA*, **89**, 3338.
19. Foote, J. and Eisen, H.N. (1995). *Proc. Natl. Acad. Sci. USA*, **92**, 1254.
20. Karlsson, R. (1994). *Anal. Biochem.*, **221**, 142.

21. Karlsson, R., Jendeberg, L., Nilsson, B., Nilsson, J., and Nygren, P-Å. (1995). *J. Immunol. Methods*, **183**, 43.
22. Butler, J.E., Ni, L., Nessler, R., Joshi, K.S., Suter, M., Rosenberg, B., *et al.* (1992). *J. Immunol. Methods*, **150**, 77.
23. Lin, J.-N., Chang, I.-N., Andrade, J.D., Herron, J.N., and Christensen, D.A. (1991). *J. Chromatogr.*, **542**, 41.
24. Nuzzo, R.G. and Allara, D.L. (1983). *J. Am. Chem. Soc.*, **105**, 4481.
25. Bain, C.D., Evall, J., and Whitesides, G.M. (1989). *J. Am. Chem. Soc.*, **111**, 7155.
26. Löfås, S. and Johnsson, B. (1990). *J. Chem. Soc. Chem. Commun.*, 1526.
27. Johnsson, B., Löfås, S., and Lindquist, G. (1991). *Anal. Biochem.*, **198**, 268.
28. Johnsson, B., Löfås, S., Lindquist, G., Edström, Å., Müller Hillgren, R.-M., and Hansson, A. (1995). *J. Mol. Recog.*, **8**, 125.
29. Löfås, S., Johnsson, B., Tegendal, K., and Rönnberg, I. (1993). *Colloids and Surfaces B: Biointerfaces*, **1**, 83.
30. Lang, H., Duschl, C., Grätzel, M., and Vogel, H. (1992). *Thin Solid Films*, **210/211**, 818.
31. Kuziemko, G.M., Stroh, M., and Stevens, R.C. (1996). *Biochemistry*, **35**, 6375.
32. Taylor, R.F. (ed.) (1991). *Protein immobilization, fundamentals and applications*. Marcel Dekker, Inc., New York.
33. Löfås, S., Johnsson, B., Edström, Å., Hansson, A., Lindquist, G., Müller Hillgren, R.-M., *et al.* (1995). *Biosens. Bioelectron.*, **10**, 813.
34. Karlsson, R., Fägerstam, L., Nilshans, H., and Persson, B. (1993). *J. Immunol. Methods*, **166**, 75.
35. Schuck, P. (1996). *Biophys. J.*, **70**, 1230.
36. Myszka, D.G., Morton, T.A., Doyle, M.L., and Chaiken, I.M. (1997). *Biophys. Chem.*, **64**, 127.
37. Karlsson, R. and Fält, A. (1997). *J. Immunol. Methods*, **200**, 121.
38. Karlsson, R. (1996). *Les Cahiers Imabio*, **17**, 29.
39. Oddie, G.W., Gruen, L.C., Odgers, G.A., King, L.G., and Kortt, A.A. (1996). *Anal. Biochem.*, **244**, 301.
40. Leatherbarrow, R.J. (1990). *Trends Biol. Sci.*, **15**, 455.
41. O´Shannessy, D.J., Brigham-Burke, M., Soneson, K.K., Hensley, P., and Brooks, I. (1993). *Anal. Biochem.*, **212**, 457.
42. Fisher, R.J. and Fivash, M. (1994). *Curr. Opin. Biotech.*, **5**, 389.
43. Roden, L.D. and Myszka, D.G. (1996). *Biochem. Biophys. Res. Commun.*, **225**, 1073.
44. Myszka, D.G., Arulanantham, P.G., Sana, T., Wu, Z., Morton, T., and Ciardelli, T.L. (1996). *Protein Sci.*, **5**, 2468.
45. Christensen, L.L.H. (1997). *Anal. Biochem.*, **249**, 153.
46. Richalet-Sécordel, P.M., Rauffer-Bruyère, N., Christensen, L.L.H., Ofenloch-Haehnle, B., Seidel, C., and van Regenmortel, M.H.V. (1997). *Anal. Biochem.*, **249**, 165.

10

Spectroscopic characterization of immobilized proteins

FRANK V. BRIGHT

1. Introduction

As a nation the United States spends billions of dollars annually on diagnostic assays. Unfortunately, most of these assays are performed in well-outfitted laboratories, and require skilled personnel, large amounts of costly reagents, and often demand long analysis times to quantify clinically important analytes. In order to overcome the disadvantages inherent with this approach, one must develop new sensing schemes that are reliable, inexpensive, fast, simple to construct and operate, redundant and self-checking, accurate, and precise, with adequate detection limits. Finally, the demand of a given sensing scheme to be used in point of care or field situations requires small, simple, and robust sensing and detection platforms.

In a generic biosensor, an immobilized biomolecule or fragment thereof serves to selectively recognize the target analyte and the binding or conversion (if the analyte is a substrate) event leads to an optical, mass, thermal, or electrochemical response that is related to the analyte concentration in the sample (1–6). Of course, there are many steps associated with the actual development of a real biosensor (1–6). For example, one must choose an appropriate biorecognition element to selectively recognize the analyte, one needs to select a detection scheme, and one must 'immobilize' the biomolecule (7–9) such that it retains its activity or affinity, it remains stable over time, and it can be reset/reused. The decision on a particular analyte: biorecognition element pair depends on the analyte of interest, its concentration in the sample, and the availability of suitable, stable biorecognition elements, and their activities or affinities. The choice of detection method depends on issues like dynamic range and requisite detection limits. Biorecognition element immobilization is *not* nearly so straightforward and, to complicate matters further, this step (often a series of steps) controls ultimately all analytical figures of merit for a given biosensor (7–9).

This chapter focuses on the use of interfacially selective fluorescence techniques to probe the behaviour of biomolecules immobilized at interfaces. The remainder of this chapter proceeds as follows: theory of total internal reflection fluorescence; theory of multifrequency phase modulation total internal reflection fluorescence (MPM-TIRF) (10, 11); instrumentation for MPM-TIRF measurements; applications of MPM-TIRF to probe the behaviour of biorecognition elements at interfaces (10–12).

2. Total internal reflection fluorescence (TIRF)

The basic principles of TIRF have been described in detail elsewhere (13–15). Briefly, consider two transparent media (1 and 2) with refractive indices n_1 and n_2 (*Figure 1*). If an incident beam of electromagnetic radiation passes through the more optically dense medium (n_1) and impinges upon the interface between n_1 and n_2 at an angle greater than the critical angle ($\Gamma_c = \sin^{-1}(n_2/n_1)$), the radiation is totally internally reflected. However, full analysis of Maxwell's wave equations (16) shows that a small portion of the incident radiation field extends into the second medium (n_2). This field is termed an evanescent wave and its $1/e$ penetration depth (into medium 2) is given by:

$$d_p(\Gamma_i) = \frac{\lambda_i}{2\pi[(n_1\sin\Gamma_i)^2 - n_2^2]^{1/2}} \quad [1]$$

where Γ_i is the angle of incidence, λ_i is the wavelength of the incident electromagnetic radiation, and the other terms are defined above. In the TIRF experiments described in this chapter, the evanescent field is used to probe those biomolecules immobilized at the 'sensor' interface.

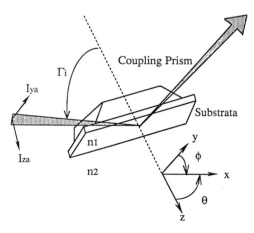

To Detector

Figure 1. Schematic of the coordinate system used in the MPM-TIRF experiment. Γ_i is the angle of incidence, I_{ya} and I_{za} are the intensity components of the excitation radiation prior to reflection, ϕ is the polar angle, and θ is the azimuthal angle.

3. Multifrequency phase and modulation (MPM) fluorescence

In the frequency domain, the sample under study is excited with high frequency (MHz-GHz) sinusoidally modulated light and the experimentally measured parameters are the frequency-dependent phase shift ($\Psi(\omega)$) and modulation ratio ($M(\omega)$) (17, 18). These values are compared by non-linear regression to the values predicted from an assumed decay law/model, and the model parameters are adjusted to yield minimal deviations between the data and the prediction. For any time domain fluorescence intensity decay law:

$$I(t) = \Sigma\alpha_i e^{-t/\tau_i} \qquad [2]$$

where $I(t)$ is the intensity at any time t, α_i is the pre-exponential factor for component i, and τ_i is the excited state fluorescence lifetime for species i. The frequency domain data are related to the sine and cosine Fourier transforms by:

$$S(\omega) = \frac{\int I(t) \sin\omega t \, dt}{\int I(t) \, dt} \qquad [3]$$

$$C(\omega) = \frac{\int I(t) \cos\omega t \, dt}{\int I(t) \, dt} \qquad [4]$$

where ω is the angular modulation frequency ($\omega = 2\pi f$, f = linear modulation frequency).

If $I(t)$ is given by *Equation 2*, the sine and cosine Fourier transforms are related to the individual model parameters (α_i and τ_i) by (17, 18):

$$S(\omega) = [\Sigma \frac{\alpha_i \omega \tau_i^2}{1 + \omega^2 \tau_i^2}] / [\Sigma\alpha_i\tau_i] \qquad [5]$$

$$C(\omega) = [\Sigma \frac{\alpha_i \tau_i}{1 + \omega^2 \tau_i^2}] / [\Sigma\alpha_i\tau_i] \qquad [6]$$

and the frequency-dependent phase angle and modulation are written (17, 18):

$$\Psi(\omega) = \arctan [S(\omega)/C(\omega)] \qquad [7]$$

$$M(\omega) = [S(\omega)^2 + C(\omega)^2]^{1/2} \qquad [8]$$

The decay terms are recovered from the experimental data by minimization of the χ^2 function:

$$\chi^2 = \frac{1}{D}\Sigma\left(\frac{\Psi(\omega) - \Psi_c(\omega)}{\delta\Psi}\right)^2 + \frac{1}{D}\Sigma\left(\frac{M(\omega) - M_c(\omega)}{\delta M}\right)^2 \qquad [9]$$

In this expression D is the number of degrees of freedom, and $\delta\Psi$ and δM are the uncertainties in the measured phase and modulation, respectively.

The subscript c denotes the frequency-dependent phase and modulation calculated based on α_i and τ_i.

To improve the precision and accuracy of the recovered model parameters, multiple sets of frequency domain data are often analysed simultaneously (when warranted) using a global analysis scheme (19–21). The average experimental phase and modulation variances are then used to minimize χ^2. In the idealized case, the best model meets simultaneously the following criteria:

(a) Simplest model having the minimum number of total floating parameters.

(b) Smallest χ^2 value.

(c) Random residual terms.

(d) Consistency with the separate experimental information.

(e) Physical significance of the chosen model.

4. Instrumentation for MPM-TIRF measurements

Figure 2 shows an abbreviated schematic of the optical interface developed for our laboratory SLM 48 000 MHF fluorometer and designed especially to study biosensor interfaces. All optical components are fused silica or calcite. The spatially inhomogeneous laser beam from the 48 000s Pockels cell modulator is directed through a beam splitter (BS); a portion directed to a reference detector and the remainder sent through the optical system. This beam is

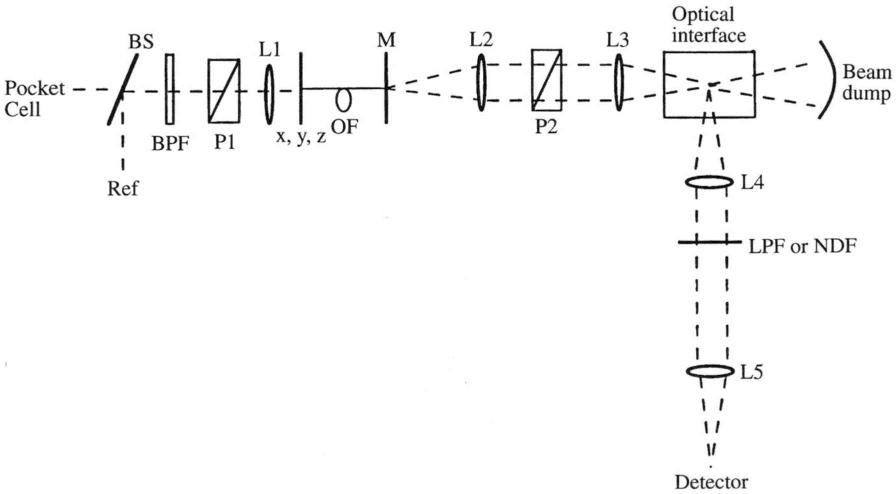

Figure 2. Schematic of the optical system used to carry out the MPM-TIRF studies of surface-immobilized biomolecules. See text for a full description of all components.

passed through a bandpass filter (BPF) to remove extraneous plasma discharge and sent through a polarizer (P$_1$) which serves as a beam attenuator. The beam is then coupled with a lens (L$_1$) into one arm of a 1000 μm core diameter optical fibre (General Fiber Optics) mounted in a precision x,y,z translation stage (Newport). A 2-m segment of optical fibre (OF) serves to effectively homogenize the excitation beam. The radiation exiting the distal end of the mounted (M) optical fibre is collected by a lens (L$_2$) and collimated, passed through a polarizer (P$_2$) set at 57° (θ_{ma}) with respect to the vertical, and focused (L$_3$) into a custom Wirth–Burbage-like optical cell designed specifically for interfacial measurements (*Figure 1*). At the n_1/n_2 interface, the focused beam is generally 1.0–1.5 mm in diameter. The excitation beam intensity is adjusted (with P$_1$) to between 3–10 μW. Exceeding 10 μW leads to significant photodegradation of the fluorescent dansyl label. Typical MPM-TIRF experiments require acquisition times of between 30–45 min.

The resulting TIRF signal from the interfacial species is collected by a f/2 lens (L$_4$) and collimated, passed through a filter (spatially homogeneous, longpass, bandpass, and/or neutral density), and focused (L$_5$) onto the photocathode of a photomultiplier tube detector. The remainder of the instrumentation for producing and detecting the high frequency modulated excitation and emission is standard 48 000 MHF.

The remainder of this chapter presents several examples using MPM-TIRF to explore the behaviour of biomolecules at interfaces. The specific examples to be described include:

(a) Probing *in situ* the distribution of free and analyte bound sites of a fluorescently labelled, silica-immobilized Fab' antibody fragment as a function of added antigen, and determination of the actual equilibrium binding constant (K_f) for the interfacial antigen:antibody complex (10). (K_f is used throughout the text to denote the binding process and *not* dissociation; $K = 1/K_f$.)

(b) Nanosecond and picosecond dynamics of acrylodan labelled bovine serum albumin physisorbed to silica as compared to the native protein in buffer (12).

4.1 Fluorescently labelled, silica-immobilized Fab' antibody fragments

The aim of these experiments is to understand the behaviour of Fab' antibody fragments immobilized on silica as a function of storage time and to determine the origin of drift seen in immunosensors based on this scheme (22–24).

Protocol 1. Preparation of fluorescently labelled, silica-immobilized Fab' fragments of anti-bovine serum IgG antibodies

Reagents

- Anti-bovine serum IgG antibodies (Sigma B 9884)
- Bovine IgG (Sigma I 5506)
- Agarose-immobilized pepsin (Sigma P 3286)
- Fluorescein labelled bovine serum albumin (Sigma A 9771)
- Human serum albumin (Sigma A 3782)
- Tris(hydroxymethyl)aminomethane (Tris) (Sigma)
- Dialysis tubing (12 000 M$_r$ cut-off) (Sigma)
- 99% pyridine (Aldrich)
- Toluene (Aldrich)
- 96% 3-glycidoxypropyltrimethoxysilane (GOPS) (Aldrich)
- 99% 2,2,2-trifluoroethane sulfonyl chloride (tresyl chloride) (Aldrich)
- 99% ethylenediaminetetraacetic acid (EDTA) (Aldrich)
- 99% dithiothreitol (Aldrich)
- Fluorescein, rhodamine 6G, perylene, and 9,10-diphenylanthracene (Aldrich)
- Acetone (Fisher)
- 48% hydrofluoric acid (Fisher)
- Nitric acid (Fisher)
- Diethyl ether (Fisher)
- Dansyl chloride (Molecular Probes)
- Fluorescein isothiocyanate (FITC) (Molecular Probes)

All other chemicals were reagent grade or better and distilled deionized water was used throughout.

Method

1. Fused silica substrates are cleaned by subjecting them to HF for 10 min, soaking in chromic acid for 1 h, and immersion in warm HNO$_3$ for 24 h.

2. Rinse cleaned substrata (step 1) with copious amounts of water, dry at room temperature, and store at room temperature in a desiccator over CaCl$_2$.

3. Silanize substrate surface by treating 10–15 quartz substrates (25 × 25 × 3 mm) with 100 ml of 10% GOPS (adjusted to pH 3.5 with phosphoric acid).

4. Heat the step 3 mixture to 90 °C for 2 h with gentle agitation.[a]

5. Rinse the substrates from step 4 with 100 ml portions of water, dry acetone, and dry ethyl ether, and dry *in vacuo* overnight.

6. Rinse the GOPS treated substrates with three washings of 50 ml dry acetone and transfer to a clean, dry beaker.

7. Add 10 ml of dry acetone and 0.7 ml of dry pyridine to the beaker with the substrates from step 6.

8. Add 200 μl of tresyl chloride slowly to the solution and swirl gently.

9. Bring the reaction mixture in step 8 to 0 °C for 25 min and wash the substrates successively with 100 ml each of 30:70, 50:50, 70:30 (acetone:5 mM HCl, v/v), and 1 mM HCl.

10. Prepare Fab' from F(ab')$_2$ (see *Protocol 4*, Chapter 3) by placing 0.1 ml of the F(ab')$_2$ solution (c.1 mg/ml total protein) into a small glass vial. To this vial add 9.9 ml of a solution composed of 0.1 M phosphate-buffered saline pH 6, 25 mM EDTA, and 6 mM dithiothreitol. Use a Pasteur pipette to bubble gently N$_2$ gas and allow the mixture to remain under these conditions for 2 h.

11. Reduce the F(ab')$_2$ disulfide bonds, form Fab', and separate the Fab' from the dithiothreitol by dialysing the mixture from step 10 against 500 ml of 0.1 M phosphate-buffered saline pH 6 containing 25 mM EDTA.

12. Maintain the entire solution in step 11 under N$_2$ and immerse the substrates from step 9 into the Fab' solution for 24 h at 4°C.

13. Wash the substrates from step 12 repeatedly with 0.1 M phosphate-buffered saline pH 7.4 and 0.1 M phosphoric acid to remove any unstable or non-specifically adsorbed antibody.

14. Rinse the immunosurfaces (step 13) with 1 mM Tris buffer pH 8.2 to inactivate the unreacted tresyl groups.

15. To block (protect) the Fab' active sites prior to fluorescent labelling, incubate the immunosurfaces with a solution containing 0.1 mg/ml bovine IgG for 10 min.

16. Rinse the substrates from step 15 with 0.1 M phosphate-buffered saline pH 7.4 and transfer to a pH 8.5 buffer solution (0.01 M) containing 0.1 μM dansyl chloride.

17. After 25–30 min at room temperature, remove the substrates in step 16 from the pH 8.5 reaction solution, rinse repeatedly with 0.1 M pH 7.4 phosphate-buffered saline and 0.1 M phosphoric acid to remove the HSA, and store in 0.1 M phosphate-buffered saline pH 7.4 at 4°C.

18. MPM-TIRF measurements[b] are carried out while the substrata is in contact with PBS at room temperature (22–25°C).

[a] It is critical to avoid Teflon stir bars here because the Teflon can be swept onto the quartz rendering it non-reactive.
[b] A detailed description of the experimental measurement protocol and data analysis schemes can be found in earlier sections or in refs 10 and 11.

One would anticipate that the fluorescent label/probe will encounter slightly different local environments in Ab* and Ag–Ab* (22–24), and the two species will undoubtedly exhibit different excited state decay kinetics. Therefore, MPM-TIRF can be used to probe directly the distribution between Ab* and Ag–Ab* at the silica interface. That is, if the fluorescent reporter group (i.e. dansyl) in Ab* and Ag–Ab* posses unique decay times, $I(t)$ will take the form:

$$I(t) = \alpha_{Ab*}e^{-t/\tau_{Ab*}} + \alpha_{Ag-Ab*}e^{-t/\tau_{Ag-Ab*}} \qquad [10]$$

Frank V. Bright

where τ_{Ab*}, $\tau_{Ag\text{-}Ab*}$, α_{Ab*}, and $\alpha_{Ag\text{-}Ab*}$ are the excited state fluorescence lifetimes associated with the free and bound forms of the antibody and the pre-exponential factors for Ab^* and $Ag\text{–}Ab^*$, respectively, and $\alpha_{Ab*} + \alpha_{Ag\text{-}Ab*} = 1$. Further, if the absorbance for the fluorescent probe (in Ab^* and $Ag\text{–}Ab^*$) are comparable, the α_x terms are related directly to the equilibrium concentration of free and antigen bound Ab^* (i.e. $[Ab^*]$ and $[Ag\text{–}Ab^*]$):

$$[Ab^*] = \alpha_{Ab}*C_{Ab}* \qquad [11]$$

$$[Ag\text{–}Ab^*] = \alpha_{Ag\text{-}Ab}*C_{Ab}* \qquad [12]$$

where C_{Ab*} is the analytical concentration of *active* antibody at the silica interface. Thus, by using MPM-TIRF one can in principle extract $\alpha_{Ab}*$ and $\alpha_{Ag\text{-}Ab}*$ which are related to $[Ab^*]$ and $[Ag\text{–}Ab^*]$ (*Equations 11* and *12*). Further substitution of the simple mass balance relationships yields K_f for the surface-immobilized Ab^*.

Figure 3 presents typical raw MPM-TIRF data for one of our anti-bovine IgG immunosurfaces in the presence of 5 nM bovine IgG (Ag) after being stored for seven days. The traces represent the best fits to single and double exponential decay laws. Attempts to fit these MPM-TIRF data to more complex models did not improve the fit quality. These data are clearly best fit by a two component model where one of the recovered lifetimes is moderately long (11 nsec) and the other short (4 nsec). These results are consistent with the presence of Ab^* and $Ag\text{–}Ab^*$. To confirm this, additional MPM-TIRF experiments were carried out as a function of added Ag (0.5–20 nM). These data were all analysed using a global analysis scheme (19–21) in which the decay times were linked and the pre-exponential factors (α_i) allowed to vary with added Ag. Again, these results demonstrate that the single exponential

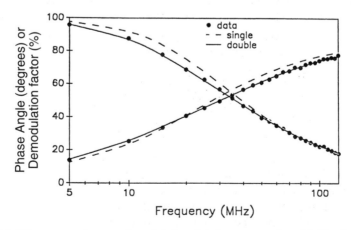

Figure 3. Multifrequency phase and modulation data for a seven-day-old silica-immobilized Ab^* preparation in the presence of 5.0 nM bovine IgG (Ag). Fits to a single (broken line) and double (solid line) exponential decay laws are shown. $\lambda_{ex} = 351.1$ nm; $\lambda_{em} > 400$ nm.

decay law fails to describe the experimental data ($\chi^2 = 1425$). Of the remaining models, the double exponential decay law best describes the observed data. Identical sets of experiments were carried out as a function of immunosurface storage time and they too were consistent with a simple two component model.

Figure 4 illustrates the recovered pre-exponential factors for the longer-lived component (τ_1) as a function of added Ag and storage time. (During the time frame of these experiments the individual excited state lifetimes remained essentially constant.) These results show several interesting trends and merit additional discussion. First, the contribution of the longer-lived component increases (symbols) and levels off as we add Ag. This is consistent with the formation of Ag–Ab*. The increase in α_1 with added Ag is due to formation of Ag–Ab* and the leveling off is a consequence of saturating all the *active* Ab* binding sites. Given this, the shorter-lived fluorescence lifetime component must represent Ab*. Secondly, the average fraction of Ag–Ab* (α_{bound}) decreases with storage time. That is, less Ag–Ab* is formed at the

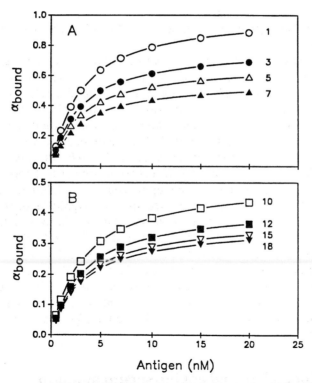

Figure 4. Pre-exponential factors for Ab*–Ag as a function of added Ag. (A) Samples stored in PBS for 1, 3, 5, and 7 days. (B) Same as (A), but storage for 10, 12, 15, and 18 days. Note: the *y* axis are different in (A) and (B).

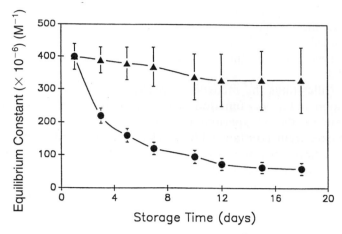

Figure 5. Effects of storage time on the recovered affinity constants. (Filled circles) Considering the total surface-immobilized Ab*. (Filled triangles) Considering only *active* surface-immobilized Ab*.

same Ag concentration when the immunosurfaces have been stored for longer periods of time.

Based on the data quality and correlation between the time resolved terms and the surface species one can calculate the average K_f for the surface-immobilized antibody (*Figure 5*). The lower trace (filled circles) illustrates the results assuming that the concentration of active silica-immobilized Ab* remains constant. Based on this assumption, one would conclude (erroneously) that K_f for Ab* decreases with storage time. However, one must realize that the analytical concentration of *active* Ab* at these silica-based immunosurfaces actually decreases with storage time (10). Therefore, the true measure of K_f is illustrated by the upper data set (filled triangles). (One should not be mislead by the differences in the magnitude of the error bars between these two traces. The relative standard deviations in K_f are essentially identical at each storage time.) Together these results demonstrate that there is essentially no change in the affinity of the surface-immobilized Fab' with time under our storage conditions. They also show that the origin of changes/drift seen in previous immunosensors based on this scheme (22–24) arises because the concentration of active antibody at the immunosurface decreases, and is *not* due to a shifting in the actual K_f for the Fab' molecules at the interface.

4.2 Acrylodan labelled bovine serum albumin

The goal of these experiments is to determine how silica adsorption influences the internal dynamics of a model protein reporter group system (BSA-Ac) (12).

Protocol 2. Preparation of acrylodan labelled bovine serum albumin

Reagents
- 6-acryloyl-(dimethylamino)-naphthalene (acrylodan, Molecular Probes)
- Essentially fatty acid-free bovine serum albumin (BSA) (Sigma)
- 12 000 M_r cut-off dialysis tubing (cellulose membrane) (Sigma)
- Na_2HPO_4 and $NaH_2PO_4.2H_2O$ (Fisher)
- Acetonitrile

All reagents were used as received without further purification, aqueous solutions were prepared in doubly distilled deionized water, and stock solutions were refrigerated in the dark at 4°C. Acrylodan was used immediately after it was dissolved in DMF.

Method

1. Prepare a 50 µM BSA stock solution in 0.1 M phosphate buffer pH 7.
2. Take 40 ml of the solution from step 1 and add enough acrylodan (in the minimum amount of acetonitrile) such that the molar ratio of BSA:acrylodan is 1:1.
3. Stir the mixture in step 2 and maintain at room temperature for 10 h.
4. Load the reaction mixture from step 3 into a 12 000 M_r cut-off cellulose dialysis bag and dialyse at 4°C against 250 ml of 1:20 (v/v) acetonitrile:phosphate buffer (0.1 M pH 7).
5. Replace the initial dialysate in step 4 after 12 h with 250 ml of 0.1 M phosphate buffer pH 7.
6. Replace the solution in step 5 every 12 h for four days.[a]
7. The final BSA-Ac solution is stored at 4°C and the molar ratio of acrylodan to BSA is generally 0.8 ± 0.05.

[a] Dialysis is complete when there is no detectable acrylodan fluorescence in the dialysate solution. In our hands, it was necessary to carry out this prolonged dialysis. Shorter-term treatments lead to significant levels (> 5%) of unreacted acrylodan in the 'BSA-Ac' solution.

Protocol 3. Adsorption of BSA-Ac to silica

Reagents
- Na_2HPO_4 and NaH_2PO_4 $2H_2O$ (Fisher)
- HNO_3 (Fisher)
- BSA-Ac from *Protocol 2*
- 1000 µm core diameter quartz multi-mode optical fibre (General Fiber Optics)
- NaOH (Fisher)

Method

1. Treat a 10 cm length of optical fibre with concentrated HNO_3 overnight to remove the cladding from a 1 cm segment.

Protocol 3. *Continued*

2. Treat the bare fibre segment from step 1 with 1 M NaOH for 30 min and wash with copious amounts of water and methanol.[a]

3. Immerse the optical fibre from step 2 into a 50 μM BSA-Ac solution (25°C). After 30 min, remove the optical fibre from this solution, rinse with 0.1 M phosphate buffer pH 7, and mount within the 48000 MHF.

[a] This process was repeated until a small drop of water spread evenly onto the bare silica and there was little evidence of cladding components in the X-ray photoelectron spectrum.

In order to follow the dynamics within the BSA-Ac system we have carried out a series of emission wavelength-dependent MPM-TIRF experiments (12) to construct the normalized time resolved emission spectra. *Figure 6* presents an abbreviated set of time-dependent emission spectra for native (upper panel), chemically denatured (central panel), and silica adsorbed (lower panel) BSA-Ac. Four features are readily apparent from *Figure 6*:

(a) The time resolved spectra for the acrylodan residue in BSA-Ac under these set of conditions are quite different from one another.

(b) The time resolved emission spectra of the acrylodan reporter group under all experimental conditions red shift following optical excitation.

(c) The emission spectra of acrylodan within native BSA-Ac is blue shifted compared to that of acrylodan within denatured BSA.

(d) The emission spectra for BSA-Ac adsorbed to silica is blue shifted relative to native and chemically denatured BSA-Ac at all times following optical excitation.

The kinetics associated with the spectral shift are a direct measure (12) of the rate of reorganization the local environment surrounding the acrylodan reporter group in BSA-Ac. The actual time-dependent spectral shift ($v(t)$) for native, chemically denatured, and silica adsorbed BSA-Ac are presented in *Figure 7* (upper panel) and reveal several interesting features. For example, there are dynamics of or surrounding the acrylodan reporter group/residue within BSA-Ac regardless of the BSA environment (i.e. native, chemically denatured, or silica adsorbed). The emission centre of gravity (i.e. weighted mean of the emission spectral profile) for the silica adsorbed BSA-Ac is greater than that for native BSA-Ac. This result is consistent with static fluorescence results (12), showing that protein adsorption induces a significant blue shift in the emission spectrum of BSA-Ac in comparison to native BSA-Ac in buffer. The total time-dependent shift of the emission for silica adsorbed BSA-Ac is larger than that seen for native BSA-Ac. Thus, it appears that silica adsorption actually increases the magnitude of the range of relaxation/dynamics surrounding the acrylodan reporter group in BSA-Ac. Compared to native BSA-Ac, the average centre of gravity shifts more

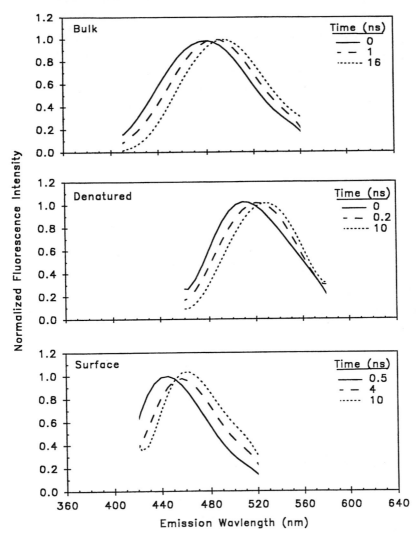

Figure 6. Time resolved emission spectra for native, chemically denatured, and silica adsorbed BSA-Ac.

rapidly for BSA-Ac at the silica surface in the first few nanoseconds; however, the spectral shift is by far the fastest in chemically denatured BSA-Ac. Finally, all dipolar relaxation is essentially completed within 4 nsec of optical excitation. This result indicates that a new equilibrium between the surrounding environment and acrylodan is reached within a 4 nsec time period.

In order to provide a more quantitative comparison of the native, chemically denatured, and silica adsorbed BSA-Ac, we calculated the dipolar relaxation function (13), $D(t)$ (*Figure 7*, lower panel). These results illustrate two

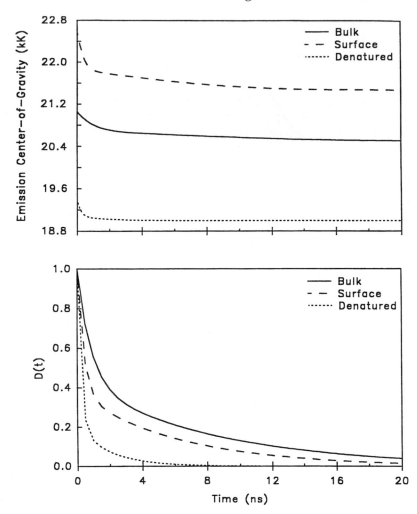

Figure 7. Time resolved emission centre of gravity (upper panel) and $D(t)$ function (lower panel) for native, chemically denatured, and silica adsorbed BSA-Ac.

key aspects of the time resolved spectral relaxation process. First, the overall rate of spectral evolution for silica adsorbed BSA-Ac is clearly faster than native BSA-Ac, but slower in comparison to chemically denatured BSA-Ac. Secondly, a single exponential decay model (one rate process) cannot accurately describe the experimental data well. A double exponential decay model (two rate processes) fits to the experimental data well and suffices to describe the spectral relaxation process. This indicates that at least *two* independent processes are responsible for the spectral relaxation of acrylodan within the BSA-Ac. These results also show that adsorption to silica strongly affects the picosecond and nanosecond dynamics of proteins at interfaces.

In summary, the results presented in this chapter show how interfacially selective time resolved fluorescence techniques (e.g. MPM-TIRF) can be used to probe and study the behaviour, performance, and dynamics of bio-molecules at interfaces under ambient conditions at less than monolayer coverages. Other techniques developed in our laboratory (e.g. phase resolved evanescent wave-induced fluorescence, PREWIF) (25) provide additional means to probe and study the behaviour of chemical recognition elements at interfaces under ambient conditions.

Acknowledgements

This work was supported in part by the National Science Foundation. I also wish to thank the many talented students from this group (current and former), whose names appear on the listed references. These individuals have contributed greatly to various aspects of our work on biorecognition element dynamics at interfaces and chemical biosensing.

References

1. Janata, J. (1989). *Principles of chemical sensors*. Plenum Press, New York, NY.
2. Wise, D. L. and Wingard, C. B., Jr. (1991). *Biosensors with fiber optics*. Humana Press, Clifton, NJ.
3. Wolfbeis, O. S. (ed.) (1991). *Fiber optic chemical sensors and biosensors*, Vol. I and II. CRC Press, Boca Raton, FL.
4. Janata, J., Josowicz, M., and DeVaney, D. M. (1994). *Anal. Chem.*, **66**, 207R.
5. Leech, D. (1994). *Chem. Soc. Rev.*, **23**, 205.
6. Taib, M. N. and Narayanaswamy, R. (1995). *Analyst*, **120**, 1617.
7. Mosbach, K. (ed.) (1987). *Methods in enzymology*, Vol. 135 and 136. Academic Press, Orlando, FL.
8. Taylor, R. F. (1991). *Protein immobilization: fundamentals and applications*, Ch. 8. Marcel Dekker, Inc., New York, NY.
9. Weetall, H. H. (1975). *Immobilized enzymes, antigens, antibodies, and peptides: preparation and characterization*, Ch. 6 and 8. Marcel Dekker, Inc., New York, NY.
10. Bright, F. V. (1993). *Appl. Spectrosc.*, **47**, 1152.
11. Bright, F. V., Wang, R., Li, M., and Dunbar, R. A. (1993). *Immunomethods*, **3**, 104.
12. Wang, R., Sun, S., Bekos, E. J., and Bright, F. V. (1995). *Anal. Chem.*, **67**, 149.
13. Axelrod, D., Burghardt, T. P., and Thompson, N. L. (1984). *Annu. Rev. Biophys. Bioeng.*, **13**, 247.
14. Andrade, J. D. (ed.) (1985). *Surface and interfacial aspects of biomedical polymers: protein adsorption*. Plenum Press, New York, NY.
15. Reichert, W. M., Suci, P. A., Ives, J. T., and Andrade, J. D. (1987). *Appl. Spectrosc.*, **41**, 503.
16. Harrick, N. J. (1967). *Internal reflection spectroscopy*. Harrick Scientific Corporation, New York.

17. Bright, F. V., Betts, T. A., and Litwiler, K. S. (1990). *CRC Crit. Rev. Anal. Chem.*, **21**, 389.
18. Gratton, E., Jameson, D. M., and Hall, R. D. (1984). *Annu. Rev. Biophys. Bioeng.*, **13**, 105.
19. Beechem, J. M., Ameloot, M., and Brand, L. (1985). *Chem. Phys. Lett.*, **120**, 466.
20. Ameloot, M., Boens, N., Andriessen, R., Van der Bergh, V., and De Schryver, F. C. (1991). *J. Phys. Chem.*, **95**, 2041.
21. Andriessen, R., Boens, N., Ameloot, M., and De Schryver, F. C. (1991). *J. Phys. Chem.*, **95**, 2047.
22. Bright, F. V., Betts, T. A., and Litwiler, K. S. (1990). *Anal. Chem.*, **62**, 1065.
23. Betts, T. A., Catena, G. C., Huang, J., Litwiler, K. S., Zhang, J., Zagrobelny, J., *et al.* (1991). *Anal. Chim. Acta*, **246**, 55.
24. Bright, F. V., Litwiler, K. S., Vargo, T. G., and Gardella, J. A., Jr. (1992). *Anal. Chim. Acta*, **262**, 323.
25. Lundgren, J. S., Bekos, E. J., Wang, R., and Bright, F. V. (1994). *Anal. Chem.*, **66**, 2433.

11

Immobilization using electrogenerated polymers

WOLFGANG SCHUHMANN

1. Introduction

The reproducible immobilization of biological recognition elements, preserving their catalytic activity and biological selectivity on the surface of a suitable transducer, is a fundamental presupposition for the successful development of biosensors. As future applications will aim at miniaturized biosensors and multisensor arrays which integrate several different biological recognition elements, immobilization procedures have to be developed which allow one:

(a) To pre-define the immobilization site (i.e. the miniaturized transducer surface) avoiding manual deposition procedures.

(b) To automate the modification of the transducer surface preferentially in a mass production compatible process (e.g. at the wafer level).

(c) To control to the maximum extent parameters influencing the reproducibility of the modification process.

Electrochemical formation of conducting polymer films occurs exclusively on the working electrode surface and hence fulfils these presuppositions to a wide extent. Several reviews summarizing this research area have been published in the last few years (1–5). However, to be able to optimize the immobilization procedure one has to become familiar—at least to a certain extent—with the parameters influencing the formation of the conducting polymer films.

2. Electrochemically-induced formation of conducting polymer films

The electrochemically-induced formation of conducting polymer films on an electrode surface involves a number of sequential steps including diffusional transport of the monomer to the electrode surface, its oxidation at an appropriate electrode potential to a radical cation, radical–radical coupling, electrochemical oxidation of the oligomers formed, chain propagation due to further coupling reactions, and finally precipitation of the polycationic polymer on the anode surface.

2.1 Mechanism of electrochemically-induced deposition of conducting polymer films

The polymerization of aromatic heterocycles, e.g. pyrrole, *N*- or 3-substituted pyrrole derivatives, thiophene and its derivatives, is initiated by the oxidation of the respective monomer at an appropriate electrode potential leading to the formation of a radical cation in the diffusion zone close to the electrode surface. The primary radical cation can react in a second step with another monomeric or oligomeric radical cation, a neutral monomer, and undergo formation of a dimeric radical cation or a dimeric diradical dication. The diradical dication may lose two protons leading to formation of the dimer (or to higher oligomers). Due to the increased delocalized π-electron system, the oxidation of these oligomers occurs in general at lower potentials as compared with the parent monomer. Hence, further radical cation formation leads predominantly to chain propagation and finally to insoluble polymer chains. After this critical chain length for precipitation is attained, the deposition of the polycationic polymer occurs on the surface of the anode. The intermediate radical cations may undergo side-reactions with nucleophilic species in the solution, preventing chain propagation. Thus, exclusion of oxygen and the proper choice of the polymerization solution is indispensable for the reproducible deposition of conducting polymer films. Local changes of the pH value due to the liberation of protons during the polymerization reaction have also to be taken into account. The resulting polymer has a net positive charge which is in general neutralized by incorporation of counter anions from the electrolyte.

2.2 Parameters influencing the electrochemical formation of conducting polymer films

Based on this complex reaction scheme, optimal conditions for the polymer formation process can be recognized. The concentration of radical cations and hence that of monomers in the diffusion zone in front of the electrode surface has to be high, the electrode potential has to be sufficiently high to allow fast electron transfer, and a solvent has to be used in which the solubility of the oligomers and polymer chains is low (6). For the reproducible formation of conducting polymer films, the polymerization potential determines—together with the temperature, the monomer concentration, the properties of the solvent, and the electrolyte—the chain length of the polymer, and thus the properties and morphology of the polymer film obtained. The thickness of the polymer film can be estimated by measuring the charge transferred during the electrochemical film formation. The electrochemical deposition can be performed by controlling the electrode potential or the current flowing through the electrode. Potentiostatic deposition regimes are recommended as the resulting diffusion profile of the monomers towards the electrode surface leads to stationary conditions and consequently to more homogeneous polymer films. Galvanostatic deposition procedures facilitate the calculation of the charge and hence the estimation of the film thickness. However, depletion of monomers in the diffusion zone around the electrode surface may lead to uncontrolled high potentials and deterioration of the film conductivity due to over-oxidation processes. A further important parameter is the choice of the electrolyte salt, i.e. the counter anion which is incorporated within the film to compensate for the net positive charge of the oxidized polymer chains. These counter anions determine to a great extent the redox properties of the conducting polymer as the nature of the counter anion defines whether, during polymer reduction, the anion can be expelled from the film or alternatively whether cations from solution have to be incorporated.

In addition to the control of these parameters, which mainly determine the properties of the conducting polymer-modified electrode, the electrode surface itself influences the homogeneity of the polymer film as well as the lateral homogeneity of the film thickness obtained. As the oxidation of the monomer is a nucleation process, the formation of radical cations occurs predominantly on active spots of the electrode surface. Consequently, the radical cation concentration is high in the vicinity of these spots, leading to a locally faster chain propagation and hence to a faster polymer growth around these areas. As a consequence, the film thickness and thus the film properties show significant lateral variations. Electrochemical platinization of the electrode surface by means of reductive deposition of small platinum crystallites from H_2PtCl_6 solution overcomes this problem (7). Reductive platinization leads to a uniform deposition of platinum clusters on the electrode surface with an average size of about 10–30 nm. As any of these platinum clusters may serve as a nucleation site and hence as a starting point for polymer deposition, the polymerization reaction takes place simultaneously over the entire electrode surface leading to a uniform lateral film thickness. In addition, due to the roughening of the electrode surface, the conducting polymer film that is formed adheres much more strongly at the platinized surface than at a polished electrode. Obviously, this technique can additionally be used to modify in a similar fashion other electrode materials such as glassy carbon, pyrolytic graphite, carbon paste, etc., and hence opens the route to easily transfer immobilization procedures from platinum to other electrodes. In *Protocol 1* the cleaning and pre-treatment of a platinum electrode is described. In *Protocol 2* the reductive platinization of an electrode is explained in detail.

Protocol 1. Electrode pre-treatment

Equipment and reagents

- Potentiostat with waveform generator; x/y recorder or computer with appropriate software for data acquisition
- Pt wire counter electrode: Hg/HgSO$_4$ reference electrode or Ag/AgCl reference electrode with a double junction diaphragm to avoid contamination of the electrolyte with Cl$^-$ ions
- Concentrated HNO$_3$, concentrated H$_2$SO$_4$
- Pt electrode set in soft glass and exposing a platinum disk of 1 mm diameter to the electrolyte
- Al$_2$O$_3$ polishing paste with grain size of 6 μm, 3 μm, 1 μm, 0.3 μm
- Polishing cloth (Technotron polishing disc MM 431, LECO)
- O$_2$-free water

Method

1. Polish the working electrode on the polishing cloth using Al$_2$O$_3$ polishing paste with decreasing grain size of 6 μm, 3 μm, 1 μm, 0.3 μm to obtain a mirror-like electrode surface.

2. Immerse the electrode in conc. HNO$_3$ for 10 min in an ultrasonic bath.

3. Rinse the electrode with water (5 min in the ultrasonic bath).

4. Wash the electrode in 10 M NaOH (10 min) and rinse with water in an ultrasonic bath (5 min).

5. Rinse with conc. H$_2$SO$_4$ for 10 min in an ultrasonic bath. Rinse with water (5 min).

6. Insert the electrode in a flask, along with a reference electrode and a Pt coil as counter electrode and remove oxygen by bubbling Ar through the electrolyte (10 min).

7. Connect the electrodes to a potentiostat and apply potential cycles at a scan rate of 100 mV sec^{-1} in 0.1 M H$_2$SO$_4$ avoiding Cl$^-$ in the electrolyte (use a Hg/HgSO$_4$ reference electrode; potential = +650 mV versus NHE or a double-junction Ag/AgCl reference electrode).

   (a) Scan from –610 to +1000 mV versus Hg/HgSO$_4$.

   (b) Scan: –810 to 1600 mV versus Hg/HgSO$_4$.

   (c) Subsequent scans –610 to +1000 mV versus Hg/HgSO$_4$ (scanning the potential between O$_2$ and H$_2$ evolution has to be continued until the cyclic voltammogram shows the separate waves for the Pt hydride formation; compare *Figure 1*).

8. Poise the electrode potential for 1 min at –210 mV versus Hg/HgSO$_4$ (as the ideal surface for the electrochemical polymerization is neither the Pt hydride nor the Pt oxide it is necessary to condition the electrode at –210 mV versus Hg/HgSO$_4$ prior to use).

9. Rinse with water.

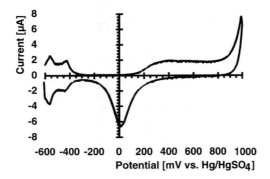

Figure 1. Typical cyclic voltammogram of a clean Pt electrode. Electrolyte: 1 M H_2SO_4, Cl⁻, and O_2-free; 1 mm diameter disk elelctrode; –600 to 1000 mV; 100 mV sec⁻¹.

Protocol 2. Platinization of the electrode surface

Equipment and reagents

- Potentiostat with waveform generator; x/y recorder or computer with appropriate software for data acquisition
- Pt wire counter electrode: Ag/AgCl reference electrode
- Cleaned Pt electrode (see *Protocol 1*) set in soft glass exposing a platinum disk of 1 mm diameter to the electrolyte
- 2 mM H_2PtCl_6 in H_2O (4 mg/ml)
- O_2-free water

Method

1. Insert the pre-treated electrode in a cell, which allows the positioning of the working electrode, a reference electrode, and a Pt coil as counter electrode, in an electrolyte volume of about 1 ml. Connect the flask via a glass valve and a vacuum tube to a high vacuum/argon line. Evacuate the flask three times to a residual pressure of 10^{-3} mbar and fill it with dry argon.

2. Transfer 2 ml of an O_2-free 2 mM H_2PtCl_6 solution in H_2O (4 mg/ml) under exclusion of O_2 into the flask (i.e. with a continuous outflow of Ar).

3. Connect the electrodes to a potentiostat and apply three potential cycles between +500 and –400 mV versus SCE at a scan rate of 10 mV sec⁻¹.

4. Remove the platinization solution from the flask by means of a Pasteur pipette and wash the electrodes three times inside the flask with O_2-free water.

5. Dry the electrodes using the vacuum line (at a final pressure of 10^{-3} mbar).

2.3 Electrochemical deposition of conducting polymer films

Based on the mechanistic considerations discussed above, limitations for the deposition of conducting polymer films on electrode surfaces can be deduced and optimized polymerization protocols can be developed. In order to understand these limitations, one has to imagine the concentration profiles of the monomer, the monomeric radical cation, as well as the higher soluble oligomers in front of the electrode surface. In the case of a potentiostatic procedure, the oxidation of the monomers occurs in front of the electrode leading to the formation of a concentration gradient and mass transport in the diffusion layer in front of the electrode surface is governed by diffusion processes. When a stationary diffusion profile is established, the concentration of the monomer at the electrode surface equals zero and the diffusion-limited current determines the maximum concentration of the primary radical cations and hence the probability of chain propagation. Assuming that the diffusion coefficients of the monomer and that of the monomeric radical cation do not differ significantly, part of the radical cations will diffuse towards the bulk of the solution and will be lost for the polymer formation. Where oligomers are formed, further chain propagation will be governed by the oxidation of these oligomers as they posses lower oxidation potentials with increased conjugation length. However, as long as the oligomers are small, they may also escape from the reaction zone to the bulk of the solution. Similar arguments lead to the description of the polymer formation process in the case of a multi-sweep deposition procedure using a triangular potential wave. Assuming that the electron transfer reaction at the electrode is fast with respect to the scan rate of the experiment, the mass transport towards the electrode surface is mainly determined by diffusion. In addition, the potential is only for a short fraction of the time at values allowing the formation of radical cations. Thus, the concentration of the primary radical cations goes through a maximum during each scan leading to a continuous variation of the probability for chain propagation and as a consequence to a variation of the film morphology.

Hence, to obtain homogeneous films and simultaneously to be able to control most of the important parameters mentioned earlier, a polymerization protocol has to be invented in which the concentration of the monomer at the electrode surface is not determined by the previously described diffusion limitations. A potentiostatic or galvanostatic pulse method for the deposition of conducting polymers on electrode surfaces has been developed (8, 9). Either the potential or the current are changed instantaneously from a value at which no monomer oxidation occurs at the electrode to a value above the monomer's oxidation potential (*Figure 2a*). The current response is the superimposition of the charging current of the electrochemical double layer and the Faradaic current due to oxidation of the monomer and subsequent oxidation of the oligomers that are formed (*Figure 2b*). Prior to the potential step, the

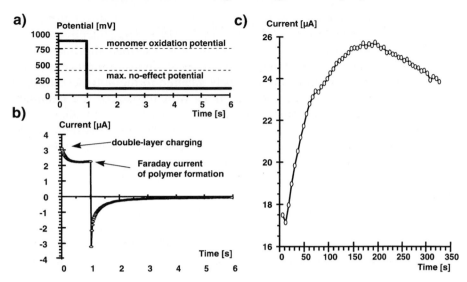

Figure 2. Pulse protocol for the electrochemical formation of conducting polymer films. (a) Potential profile of one pulse of a sequence for the deposition of polypyrrole. The potential is changed by an immediate jump from a value beneath a no-effect potential above the oxidation potential of the monomer. After a defined deposition time (here: 1 sec), the potential is again changed back and held beneath the no-effect value at which the restoration of bulk concentrations in front of the electrode takes place (here: 5 sec). (b) Current response obtained with the potential pulse from (a). After the charging of the double layer the current decreases to a value which is determined by the complex oxidation reactions at the electrode. (c) Current values at the very end of the deposition time plotted over the time of the experiment. Each point represents the current at the end of one potential pulse. The current increase due to the increase of the electrode surface at the beginning of the experiment clearly proves the successful deposition of the conducting polymer film. With increasing film thickness, further polymer deposition becomes more difficult due to the smaller conductivity of the formed polymer as compared with the bare electrode.

monomer concentration in front of the electrode is equal to its bulk concentration and is independent of mass transport limitations. Depending on the geometry of the electrode and the rate of the oxidation process, which determines the time necessary to establish the diffusion profile, the potential is changed back after a predefined pulse duration either to an intermediate value at which oligomer oxidation is possible or directly to the starting potential. During a resting phase, the bulk concentration of the monomer is reestablished in front of the electrode surface by diffusional mass transport and simultaneously soluble side-products are able to diffuse away from the reaction zone. This sequence is repeated until the required polymer film thickness (determined by measuring the charge transferred during the electrochemical deposition process) is obtained. Usually, the current is only recorded at the very end of the pulse to discriminate for the double layer charging current

(*Figure 2c*). Thus, the growth of the film can be directly visualized during the electrochemical polymerization process. In *Protocol 3* the electrochemical deposition of polypyrrole from aqueous solution is described alternatively for a multi-sweep deposition and a potentiostatic pulse deposition procedure. As a matter of fact, this protocol can be easily adapted to the polymerization of other monomers by adjusting the potentials with respect of the oxidation potential of the specific monomer.

Protocol 3.   Electrochemical deposition of polypyrrole

Equipment and reagents

- Potentiostat with waveform generator or computer with digital-to-analog conversion board; x/y recorder or computer with appropriate software for data acquisition
- Pt wire counter electrode; Ag/AgCl reference electrode
- Cleaned and platinized Pt electrode
- KCl: 100 mM solution in H_2O

- Pyrrole: best available purity; at least 99%
- Glass column: length 5 cm, diameter 0.4 cm, packed with neutral alumina (Al_2O_3)
- Electrochemical cell consisting of a three-necked flask with glass valve for connection to a high vacuum/argon line
- O_2-free water

Method

1. Insert the pre-treated electrode in a flask, which allows the working electrode to be positioned, together with a reference electrode, and a Pt coil as counter electrode in an electrolyte volume of about 1 ml. Connect the flask via a glass valve and a vacuum tube to a high vacuum/argon line, evacuate the flask three times to a residual pressure of 10^{-3} mbar, and fill it with dry Ar.

2. Purify pyrrole by passing 0.5 ml pyrrole (as received) through a neutral Al_2O_3 column (length 5 cm, diameter 0.4 cm) to remove any coloured components.

3. Dissolve 56 μl pyrrole in 1.944 ml of 100 mM KCl rigorously excluding any O_2 (final concentration of pyrrole is 100 mM). Transfer this solution into the electrochemical cell with a continuous outflow of Ar.

4. Apply either:

   (a) A potentiostatic pulse profile to the working electrode with potentials of +875 mV for 1 sec (deposition phase) and 0 mV for 5 sec (resting phase) (see *Figure 2a–c*).

   (b) A triangular potential wave (–300 mV to +875 mV versus Ag/AgCl; 100 mV sec^{-1}).

   The film thickness is determined by the number of deposition pulses or the number of triangular potential scans, respectively.

---

3. Immobilization of biological selectivity elements in conducting polymer films

The formation of modified electrode surfaces by means of electrochemical polymerization procedures can be advantageously applied to the immobilization of biological recognition elements on electrode surfaces. In principle, there are two different approaches to immobilize enzymes or other biological selectivity elements using the electrochemically-induced formation of conducting polymers. The first and most widely used is to entrap the enzyme within the growing ramified polymer network during its electrochemical formation (10–14). The second extends this procedure to modified biological recognition elements which are covalently bound to the polymer chains during their entrapment. The third uses a two-step procedure consisting of the formation of a functionalized conducting polymer film followed by the covalent binding of the biocatalysts to the functionalities at the polymer surface (15). Due to the generally large size of biological recognition elements as compared with the pore size of the polymer, the biomolecules are in this case exclusively bound to the outer polymer surface while they should be evenly distributed within the film following the first and second approach (*Figure 3*).

In the following subsections, these different immobilization strategies will be described in detail using as examples the entrapment of glucose oxidase

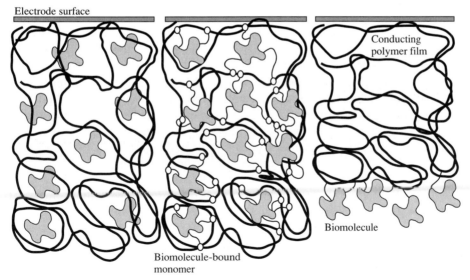

Figure 3. Schematic representation of immobilization strategies for biomolecules using conducting polymer films. Left: polymer entrapped biomolecule. Middle: covalent entrapment of biomolecules using monomer-modified biological recognition elements. Right: two-step modification leading to covalent binding of the biomolecule at the outer surface of the functionalized conducting polymer film.

within a growing polypyrrole film, the covalent entrapment of pyrrole-modified glucose oxidase in polypyrrole films, and the covalent binding of glucose oxidase at an amino functionalized polypyrrole film.

3.1 Entrapment of biological selectivity elements in conducting polymer films (one-step immobilization)

For the one-step immobilization of biomolecules in growing conducting polymer films, there are severe restrictions as the biomolecule has to be dissolved in the electrolyte solution together with the monomer while its biological activity has to be preserved. Consequently, this procedure is limited to monomers which:

- can be dissolved in aqueous solution (e.g. pyrrole, pyrrole derivatives)
- have an oxidation potential sufficiently smaller than the oxidation potential of water.

In addition, the pH value of the deposition solution is limited by the properties of the biomolecule and has to be controlled during the film formation procedure.

In order to optimize the entrapment of large biomolecules in the growing ramified polymer network, one again has to imagine the complex reaction sequence of the electrochemically-induced polymer formation but now in the presence of the biomolecule. As the diffusion of the high molecular weight biomolecules to the electrode is slow as compared with the diffusion of the monomers, the number of entrapped biological selectivity elements is mainly limited to those which are already close to the reaction zone prior to the start of the electrochemical polymerization process. Consequently, one would assume that the concentration of the biomolecule has to be high in the deposition solution in order to augment the immobilized biological activity within the conducting polymer film. However, although increased immobilized enzyme activity has been observed with increasing concentration of glucose oxidase in the deposition solution (11) the nucleophilic side chains at the biomolecule interfere—as has been already pointed out above—with the polymer growth due to the attack on the intermediate radical cations. Hence, at high enzyme concentrations either film formation fails or the deposition time is long leading to films with dramatically changed morphology (9).

As an alternative, the pulse deposition procedure allows enhancement of the immobilized enzyme activity while keeping the concentration of the enzyme low. The resting phase prior to the next potentiostatic pulse has to be long enough to allow re-establishment of both the bulk concentration of the monomer and that of the biomolecule at the electrode surface. Due to the different numbers of nucleophilic groups at the outer surface of biomolecules the procedure has to be optimized for different biological recognition elements by varying the pulse length, the monomer concentration, the biomolecule concentration in the deposition solution, and the ionic strength of the electrolyte used. The method may be extended to even larger structures, e.g. whole living cells for their entrapment in the growing polymer film. In *Protocol 4* the entrapment of glucose oxidase in a polypyrrole film is described in detail. Although the pulse deposition protocol has significant advantages, the multi-sweep deposition procedure is given as an alternative.

Protocol 4. Entrapment of glucose oxidase in a growing
polypyrrole film

Equipment and reagents

- Potentiostat with waveform generator or computer with digital-to-analog conversion board; x/y recorder or computer with appropriate software for data acquisition
- Pt wire counter electrode: Ag/AgCl reference electrode
- Cleaned and platinized Pt electrode
- KCl: 100 mM solution in H_2O
- Pyrrole: best available purity; at least 99%

- Al_2O_3 column: neutral, length 5 cm, diameter 0.4 cm
- Electrochemical cell consisting of a three-necked flask with glass valve for connection to a high vacuum/argon line
- O_2-free water
- Glucose oxidase from *Aspergillus niger* (Sigma, Type X)

Method

1. Insert the cleaned and platinized electrode in a flask, which allows for positioning of the working electrode, a reference electrode, and a Pt coil as counter electrode in an electrolyte volume of about 1 ml. Connect the flask via a glass valve and a vacuum tube to a high vacuum/argon line, evacuate the flask three times to a residual pressure of 10^{-3} mbar, and fill it with dry Argon.

2. Purify the pyrrole by passing 0.5 ml pyrrole (as received) through a neutral Al_2O_3 column to remove any coloured components.

3. Dissolve 56 μl pyrrole in 1.944 ml of 100 mM KCl rigorously excluding any O_2 (final concentration of pyrrole is 100 mM).

4. Add 2 mg glucose oxidase and transfer this solution into the electro-chemical cell under a continuous outflow of Ar.

5. Apply either:

   (a) A potentiostatic pulse profile to the working electrode with potentials of +875 mV for 1 sec (deposition phase) and 0 mV for 10 sec (resting phase).

   (b) A triangular potential wave (−300 mV to +875 mV versus Ag/AgCl; 100 mV sec^{-1}).

3.2 Covalent entrapment of biological selectivity elements in conducting polymer films

As has been pointed out, the entrapment of a biomolecule during the electro-chemical formation of a conducting polymer layer is mainly due to a statistical enclosure. In addition, positively charged biomolecules having an isoelectric point higher than the pH value of the deposition solution are usually hard to integrate into the growing polymer layer as they are expelled by the positive polymer chains to the bulk of the solution. A possible approach is based on the modification of the biomolecule with a functionalized monomer followed by the electrochemically-induced copolymerization of the monomer-modified biomolecule with the parent monomer leading to a covalent linkage between the polymer entrapped biological recognition element and the conducting polymer chains.

Until now most work has been done using glucose oxidase and pyrrole as model compounds. Pyrrole-modified glucose oxidase can be obtained by means of various condensation reactions between functionalized pyrrole derivatives and corresponding functionalities at the surface of the enzyme (*Figure 4*). The

(a) In aqueous solution

Monomer (e.g. pyrrole)

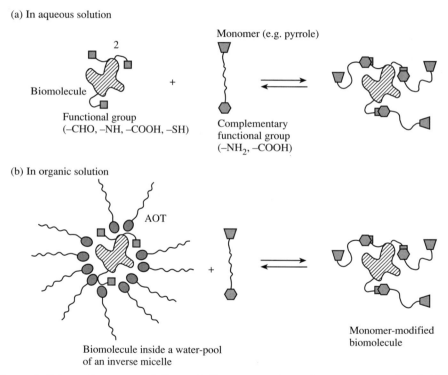

Biomolecule

Functional group
(–CHO, –NH, –COOH, –SH)

Complementary
functional group
(–NH₂, –COOH)

(b) In organic solution

AOT

Biomolecule inside a water-pool
of an inverse micelle

Monomer-modified
biomolecule

Figure 4. Schematic representation of the preparation of monomer-modified biomolecules. (a) Condensation of functional side-groups at the enzyme with complementary functionalized monomers (e.g. formation of Schiff bases, amides, thioesters) in aqueous solution. (b) Condensation of functional side-groups at the surface of the biomolecule with water-insoluble complementary functionalized monomers using a phase transfer reaction. Deactivation of the biomolecules is prevented by their inclusion in the water-pool in the interior of inverse micelles. In (a) the functionalized monomer has to be soluble in aqueous solution, in (b) in organic solvents.

formation of amide bonds between carbodiimide-activated *N*-(2-carboxyethyl) pyrrole and lysyl residues at the surface of glucose oxidase leads to an average of 30 pyrrole units per enzyme molecule which also has a stabilizing effect on the enzyme (16, 17). Similarly, *N*-(3-aminopropyl)pyrrole has been bound to carbodiimide-activated carboxylic residues at the enzyme (18). A third possibility uses aldehyde functions at the enzyme obtained by means of IO_4^- oxidation of the oligosaccharide chains of the glycoprotein glucose oxidase. The covalent binding of *N*-(ω-aminoalkyl)pyrrole derivatives (15, 19) as pyrrole components involves the primary formation of Schiff bases at the surface of the enzyme, which can be subsequently reduced to the more stable secondary amines using $NaBH_4$. Obviously, the periodate oxidation of vicinal diols is only applicable for glycoproteins bearing a sufficient number of sugar residues on their surface. The reaction conditions, especially the periodate concentration, the reaction time, and the reaction temperature have to be optimized for each individual glycoprotein in order to obtain a high number of aldehyde groups and minimal loss of biological activity. In *Protocol 5* the synthesis of *N*-substituted pyrrole derivatives with terminal amino functions or terminal carboxylic acid groups is described. *Protocol 6* describes the oxidation of sugar residues at the surface of the glycoprotein glucose oxidase, and in *Protocol 7* the modification of either the native enzyme with carboxylic acid functionalized pyrrole monomers or the oxidized enzyme with amino functionalized pyrrole monomers is described in detail. Similar enzyme modification can in principal be done with 3-substituted pyrrole derivatives (for synthesis of such compounds see refs 20–22).

Protocol 5. Synthesis of functionalized pyrrole derivatives

Equipment and reagents

- Three-necked flask with reflux condenser and dropping funnel
- Magnetic stirrer with heating plate and oil-bath
- Separation funnel
- Equipment for high vacuum distillation
- Acetic acid
- 1,6-diaminohexane (or alternatively 12-aminododecanoic acid)
- 2,5-dimethoxytetrahydrofuran
- CH_3Cl
- Na_2SO_4 (anhydrous)

Method

1. Prepare a mixture of 300 ml H_2O and 200 ml acetic acid.

2. Dissolve 5.8 g 1,6-diaminohexane in 375 ml of the H_2O:acetic acid mixture in a 1 litre three-necked flask with dropping funnel and reflux condenser (for the carboxylic acid terminated compound use 10.75 g of 12-aminododecanoic acid).

3. Stir with a magnetic stirring bar and heat to reflux.

4. Dissolve 6.5 ml 2,5-dimethoxytetrahydrofuran in 125 ml of the

Protocol 5. *Continued*

H_2O:acetic acid mixture. Add this solution dropwise (over a period of at least 2 h) to the boiling reaction mixture using the dropping funnel.

5. Heat to reflux for an additional hour and then allow the reaction mixture to cool to room temperature.

6. Extract the reaction mixture with 100 ml CH_3Cl (three times) using a separation funnel.

7. Dry the organic phase over Na_2SO_4; remove most of the solvent under reduced pressure.

8. Purify the dark brown oil by means of high vacuum distillation leading to a colourless product of *N*-(6-aminohexyl)pyrrole or *N*-(12-carboxy-dodecyl)pyrrole, respectively.

Protocol 6. Introduction of aldehyde groups at the surface of glucose oxidase by oxidation of sugar residues with periodate

Equipment and reagents

- Thermostatted water-bath
- Flask with heat exchanger
- Ultrafiltration cell with membranes of 5000 or 10 000 Da M_r cut-off
- Column for size exclusion chromatography: 2.5 cm diameter, 50 cm length, filled with Sephadex G25 coarse

- Peristaltic pump to adjust a continuous flow of the eluent through the column
- $NaHCO_3$
- $NaIO_4$
- Ethylene glycol
- Glucose oxidase from *Aspergillus niger* (Sigma, Type X)

Method

1. Dissolve 100 mg glucose oxidase in 10 ml of a 600 mM $NaHCO_3$ pH 8.1, and adjust the temperature to 25°C.

2. Add 20 ml of a 50 mM $NaIO_4$ solution and let the reaction take place for exactly 30 min with gentle shaking.

3. Add 20 ml of a 50 mM ethylene glycol solution in H_2O to stop the reaction and continue for a further hour with gentle shaking.

4. Concentrate the reaction mixture to 5 ml by means of ultrafiltration through an ultrafiltration membrane with a M_r cut-off of 5000–10 000 Da, and wash several times with 10 mM $NaHCO_3$.

5. Separate the enzyme from low molecular weight compound by means of size exclusion chromatography (Sephadex G25 coarse; water as eluent). Collect the enzyme-containing yellow fractions. Freeze them in portions of 1 ml or remove the water from the modified enzyme by freeze drying (this reduces the enzyme activity by about 30%).

Protocol 7. Modification of glucose oxidase with functionalized pyrrole derivatives

Equipment and reagents

- Ultrafiltration cell with membranes of 1000 Da M_r cut-off
- Column for size exclusion chromatography: 2.5 cm diameter, 50 cm length, filled with Sephadex G25 coarse
- Peristaltic pump to adjust a continuous flow of the eluent through the column
- 100 mM phosphate buffer pH 7
- $NaBH_4$
- Glucose stock solution in phosphate buffer (allowed to anomerize overnight)

- Functionalized pyrrole derivative
- CH_3CN
- Dicyclohexylcarbodiimide
- *N*-hydroxysuccinimide
- Dioctylsodiumsulfosuccinate (AOT)
- Octane
- Borate buffer
- Acetone
- Periodate oxidized glucose oxidase or glucose oxidase from *Aspergillus niger* (Sigma, Type X)

A. *Condensation of amino functionalized pyrrole derivatives with periodate oxidized glucose oxidase*

1. Dissolve 100 mg of periodate oxidized glucose oxidase (see *Protocol 6*) in 120 ml of 100 mM phosphate buffer pH 7 containing 1 mM glucose.

2. Add 50 mg of the amino functionalized pyrrole derivative (e.g. *N*-(6-aminohexyl)pyrrole) (see *Protocol 5*) and allow to react with gentle shaking overnight in a cold room at 4°C.

3. Add 500 mg $NaBH_4$ and shake gently for 1 h at room temperature to reduce the Schiff bases.

4. Concentrate the reaction mixture to 5 ml by means of filtration through an ultrafiltration membrane with a M_r cut-off of 1000 Da and wash several times with 100 mM phosphate buffer pH 7.

5. Separate the enzyme from low molecular weight compounds by means of size exclusion chromatography (Sephadex G25 coarse; water as eluent). Collect the enzyme-containing yellow fractions. Freeze them in portions of 1 ml (freeze drying reduces the enzyme activity significantly).

B. *Condensation of carboxy functionalized pyrrole derivatives with native glucose oxidase using reverse micelles (23)*

1. Dissolve 25 mg of the carboxy modified pyrrole derivative (e.g. *N*-(5-carboxypentyl)pyrrole) (see *Protocol 5*) in 500 μl CH_3CN, and add a solution of 30 mg dicyclohexylcarbodiimide in 500 μl CH_3CN. Allow the urea derivative to precipitate and add 15 mg of *N*-hydroxysuccinimide. Shake the reaction mixture at room temperature for 30 min (solution A).

2. Dissolve 27 g AOT in 200 ml octane (solution B).

3. Dissolve 100 mg native glucose oxidase in 25 ml 50 mM borate buffer pH 9 (solution C).

Protocol 7. *Continued*

4. Cool solution B and C in an ice-bath. Add solution C slowly to solution B, and stir with a glass stick until a clear solution with no phase separation is obtained.

5. Add solution A to the reaction mixture from step 4 and allow to react under gentle shaking at room temperature for at least 1 h (the reaction mixture can be kept overnight in the refrigerator).

6. Add 250 ml ice-cold acetone to precipitate the modified enzyme and collect it immediately by filtration through a paper filter.

7. Dissolve the enzyme in 100 mM phosphate buffer pH 7 and remove low molecular weight contaminants by means of size exclusion chromatography (Sephadex G25 coarse; water as eluent). Collect the enzyme-containing yellow fractions. Freeze them in portions of 1 ml (freeze drying reduces the enzyme activity significantly).

Note: it is also possible to link amino functionalized pyrrole derivatives to carboxylic acid residues at the surface of the biomolecule after their activation with a water soluble carbodiimide (compare *Protocol 9*).

The glucose oxidase bound pyrrole units can be electrochemically oxidized at an electrode surface with an oxidation potential similar to that of free *N*-substituted pyrrole derivatives. However, as expected from the size of an individual enzyme molecule, homopolymerization of pyrrole-modified glucose oxidase could not be observed. Therefore, the formation of copolymers between free and enzyme-bound pyrrole is necessary. As the oxidation of the enzyme-linked pyrrole moieties occurs at higher potentials than the oxidation of the parent monomer, in a potentiostatic, galvanostatic, or multi-sweep deposition process, the integration of enzyme-linked pyrrole moieties into the polymer film is disfavoured. However, by application of a pulse deposition procedure, the covalent entrapment of pyrrole-modified glucose oxidase is mainly determined by its concentration in the vicinity of the electrode. Although no significant steric hindrance for the oxidation of pyrrole units bound via short spacer chains to the enzyme's surface have been observed, it can be assumed that long and flexible spacer chains may improve the probability of the enzyme-linked pyrrole being integrated into the growing polymer chain. As the pyrrole-modified enzyme seems to be stabilized after integration into the copolymer film most likely by decreasing its mobility within the polymer network, this approach should be extended to other biological recognition elements.

Despite the much more difficult synthesis of pyrrole derivatives functionalized in the 3-position, their covalent binding to the surface of biomolecules seems to exhibit at least two advantages over the use of *N*-substituted pyrrole derivatives (24). First, related polypyrrole films should have a higher conductivity which should facilitate the formation of thick films, and secondly, the oxidation potentials of pyrrole and 3-substituted pyrroles should be more similar. Thus, the composition of the copolymer film can be more easily controlled via the concentration ratio of pyrrole and enzyme bound pyrrole units leading to an improved immobilized enzyme activity within the copolymer film.

Protocol 8.   Copolymerization of pyrrole-modified glucose oxidase and pyrrole

Equipment and reagents

- Potentiostat with waveform generator or computer with digital-to-analog conversion board; x/y recorder or computer with appropriate software for data acquisition
- Pt wire counter electrode: Ag/AgCl reference electrode
- Cleaned and platinized Pt electrode (see *Protocols 1* and *2*)
- KCl: 100 mM solution in H$_2$O

- Pyrrole: best available purity; at least 99%
- Glass column: length 5 cm, diameter 0.4 cm, packed with neutral alumina (Al$_2$O$_3$)
- Electrochemical cell consisting of a three-necked flask with glass valve for connection to a high vacuum/argon line
- O$_2$-free water
- Pyrrole-modified glucose oxidase (see *Protocol 7*)

Method

1. Insert the pre-treated electrode in a flask, which allows positioning of the working electrode, a reference electrode, and a Pt coil as counter electrode in an electrolyte volume of about 1 ml. Connect the flask via a glass valve and a vacuum tube to a high vacuum/argon line, evacuate the flask three times to a residual pressure of 10^{-3} mbar, and fill it with dry argon.

2. Purify pyrrole by passing 0.5 ml pyrrole (as received) through a neutral Al$_2$O$_3$ column (5 cm × 0.4 cm) to remove any coloured components.

3. Dissolve 14 μl pyrrole in 1.986 ml of 100 mM KCl containing 10 mg of the pyrrole-modified enzyme rigorously excluding O$_2$ (final concentration of pyrrole is 25 mM). Transfer the solution into the electrochemical cell und a continuous outflow of Ar.

4. Apply a potentiostatic pulse profile to the working electrode with potentials of +1100 mV for 1 sec (deposition phase) and 0 mV for 10 sec (resting phase). The film thickness is determined by the number of deposition pulses.

3.3 Covalent immobilization of biological selectivity elements at functionalized conducting polymer films (two-step immobilization)

A second possibility for the immobilization of enzymes at conducting polymer-modified electrode surfaces makes use of functionalized conducting polymer films as partners for the formation of covalent bonds with biological recognition elements. Either an already functionalized monomer (see *Protocol 5*) can be polymerized or copolymerized with the unsubstituted parent monomer under direct formation of the functionalized surface or an unsubstituted polymer film can be derivatized in a heterogeneous reaction on the electrode surface after the formation of the conducting polymer film (*Figure 5*). These functionalized electrode surfaces can be subsequently used for covalent binding of suitable biological selectivity elements in a second step. This sequential procedure has obvious advantages over a one-step entrapment process due to the inherent possibility of choosing the optimum reaction conditions for each single step of the procedure. The conducting polymer film can be grown from organic solvents, functionalization may be performed under exclusion of water and oxygen, and finally it is possible to switch to an environment which is best suited for the covalent binding of the envisaged biomolecule. In particular, the formation of the polymer under conditions which may be deleterious for biological compounds expands significantly the number of potentially applicable conducting polymers. In addition, labile biomolecules which may

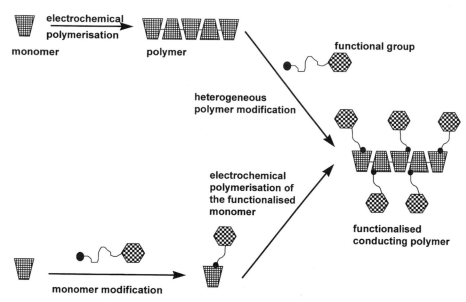

Figure 5. Possibilities for the formation of functionalized conducting polymer films. Top: modification of the already deposited conducting polymer according to a heterogeneous reaction (e.g. nitration of polypyrrole films). Bottom: synthesis of functionalized monomers and their subsequent electrochemical polymerization under formation of the functionalized conducting polymer film. Here, the modification of the polymer is more uniform, however, there are sometimes problems with the polymerization of these monomers especially when the functionalized groups are bulky or interfere with the chain propagation due to a nucleophilic character.

be deactivated during entrapment due to uncontrolled low pH values near the electrode surface may be used in conducting polymer-based biosensors.

One possibility for the derivatization of already grown polypyrrole films in a heterogeneous reaction involves the nitration of polypyrrole using *in situ* generated acetyl nitrate and the subsequent electrochemical reduction of the pyrrole-bound nitro groups with the formation of poly(3-aminopyrrole) (25–27). Glucose oxidase has been immobilized via amide bonds generated from activated carboxylic side chains of the enzyme and the polymer-linked amino functions. As expected from the location of the enzyme at the outer surface of the polymer film, the response characteristics of these electrodes are dependent on the thickness of the polymer layer. The surface area and thus the amount of immobilized enzyme activity increases with increasing polymer film thickness while the diffusion distance for H_2O_2 gets larger leading to a lower probability of a H_2O_2 molecule reaching the electrode surface.

Another approach to obtain functionalized conducting polymer films for covalent attachment of biomolecules is based on the polymerization of *N*-substituted pyrrole derivatives (for synthesis see *Protocol 5*) and subsequent covalent binding of the biocompound (28, 29). Similar modified electrodes have been obtained using amino functionalized azulene derivatives (30), amino functionalized dithienylpyrrole derivatives (31), and functionalized thiophene or bithiophene derivatives (32). The sequential two-step process allows tailoring of the properties of the underlying conducting polymer film before binding the enzyme. As the porosity and morphology of the polymer films can be varied by adjusting proper deposition conditions, the inherent size exclusion properties of the polymer film can be used to prevent interfering compounds from being oxidized at the electrode surface (33, 34). In *Protocol 9* two approaches to the formation of amino-modified polypyrrole films are described. The first is based on the derivatization of a polypyrrole film by means of a heterogeneous nitration reaction, followed by electrochemical reduction to the amino functions, and the covalent binding of glucose oxidase. The second uses the direct electrochemical polymerization of amino functionalized pyrrole derivatives and the subsequent covalent binding of the biological recognition element.

Protocol 9. Covalent immobilization of glucose oxidase at an amino functionalized polypyrrole film

Equipment and reagents

- Potentiostat with waveform generator or computer with digital-to-analog conversion board; x/y recorder or computer with appropriate software for data acquisition
- Pt wire counter electrode: Ag/AgCl reference electrode
- Cleaned and platinized Pt electrode

- CH_3CN: distilled and stored over molecular sieve 4 Å under argon in the dark
- Tetrabutylammonium *p*-toluenesulfonate (TBATos)
- 1-cyclohexyl-3-(2-morpholinoethyl)carbodiimide-metho-*p*-toluenesulfonate (CCD)
- Acetic anhydride

Protocol 9. *Continued*

- $Cu(NO_3)_2 \cdot 3H_2O$
- Glucose stock solution (allowed to anomerize overnight prior to use)
- Pyrrole: best available purity; at least 99%, or a suitable *N*-(ω-aminoalkyl)pyrrole derivative (for synthesis see *Protocol 5*)
- Glass column: length 5 cm, diameter 0.4 cm, packed with neutral alumina

- Electrochemical cell consisting of a three-necked flask with glass valve for connection to a high vacuum/argon line
- O_2-free water
- Glucose oxidase from *Aspergillus niger* (Sigma, Type X) or periodate oxidized glucose oxidase (for preparation see *Protocol 6*)

Method

1. Insert the cleaned and platinized electrode in a flask, which allows positioning of the working electrode, a reference electrode, and a Pt coil as counter electrode in an electrolyte volume of about 1 ml.

2. Connect the flask via a glass valve and a vacuum tube to a high vacuum/argon line, evacuate the flask three times to a residual pressure of 10^{-3} mbar, and fill it with dry argon.

3. (a) Purify pyrrole by passing 0.5 ml pyrrole (as received) through a neutral Al_2O_3 column to remove any coloured components. Dissolve 56 μl pyrrole in 1.944 ml of 100 mM CH_3CN (final concentration of pyrrole is 100 mM) containing 100 mM TBATos rigorously excluding any O_2 (continue with step 4a).

   (b) Alternatively dissolve a suitable *N*-(ω-aminoalkyl)pyrrole derivative (see *Protocol 5*) (final concentration of the pyrrole derivative is 100 mM) in CH_3CN or CH_2Cl_2 containing 100 mM TBATos (continue with step 4b).

4. (a) Apply a potentiostatic pulse profile to the working electrode with potentials of +875 mV for 1 sec (deposition phase) and 0 mV for 5 sec (resting phase) (continue with step 5).

   (b) Apply a potentiostatic pulse profile to the working electrode with potentials of +1200 mV for 1 sec (deposition phase) and 0 mV for 5 sec (resting phase) (continue with step 8).

5. Nitrate the polypyrrole film by immersing the electrode into a solution of 700 mg $Cu(NO_3)_2.3H_2O$ in 20 ml acetic anhydride for 5 min at 20 °C under an argon atmosphere.

6. Rinse the electrode three times with O_2-free CH_3CN.

7. Reduce the nitro groups formed electrochemically by means of three potential cycles between +500 mV and −2500 mV versus SCE with a scan rate of 10 mV sec^{-1} in a CH_3CN solution containing 100 mM TBATos.

8. Activate the carboxylic side chains of 10 mg/ml glucose oxidase using a 100 mM CCD in 100 mM buffer (pH 4.5) containing 100 mM glucose.

9. Immerse the modified electrode surface for 3 h in the activated enzyme solution.

10. Rinse the electrode surface extensively with 100 mM KCl to remove any enzyme that is only adsorbed.

4. Immobilization of different biomolecules on the surface of individual electrodes of an electrode array

As was pointed out in the introduction of this chapter, one major advantage of the electrochemical production of conducting polymers over deposition techniques necessary for non-conducting polymers is the exact localization of the polymerization process. The electrochemical polymerization process is exclusively localized at the electrode surface to which the appropriate deposition potential is applied, and hence, this polymer deposition is independent the size and/or geometry of the electrode surface. Even microelectrodes with a non-planar geometry can be precisely modified with an electrochemically deposited polymer, obtaining layers with a rather homogeneous thickness from some nanometres to several micrometres. This opens a unique route to modify individual electrodes in an electrode array with different biological recognition elements (*Figure 6*). Using the one-step entrapment method described in Section 3.1 and *Protocol 4* sequentially is one possible approach for this individual electrode modification (35) which is described in detail in

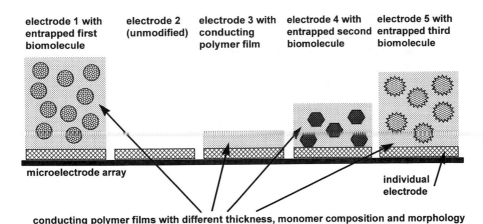

electrode 1 with entrapped first biomolecule electrode 2 (unmodified) electrode 3 with conducting polymer film electrode 4 with entrapped second biomolecule electrode 5 with entrapped third biomolecule

microelectrode array

individual electrode

conducting polymer films with different thickness, monomer composition and morphology

Figure 6. Schematic representation of the individual modification of electrodes from a microelectrode array using conducting polymers. Different biomolecules can be entrapped, different film thickness and film morphology can be chosen to optimize the response characteristics of the individual specific sensors.

Protocol 10. Additionally, using this technique there are possibilities for the integration of anti-interference layers or multiple enzyme electrodes with different linear detection ranges in one electrode array.

Protocol 10. Sequential modification of individual electrodes of an array by means of entrapment of the biomolecules in growing polypyrrole films

Equipment and reagents

- Potentiostat with waveform generator or computer with digital-to-analog conversion board; x/y recorder or computer with appropriate software for data acquisition
- Pt wire counter electrode: Ag/AgCl reference electrode
- Cleaned and platinized electrode array
- KCl: 100 mM solution in H_2O
- Pyrrole: best available purity; at least 99%

- Glass column: length 5 cm, diameter 0.4 cm, packed with neutral alumina
- Electrochemical cell consisting of a three-necked flask with glass valve for connection to a high vacuum/argon line
- O_2-free water
- Glucose oxidase from *Aspergillus niger* (Sigma, Type X)

Method

1. Insert the cleaned and platinized electrode array in an electrochemical cell with small electrolyte volume and the possibility of removing O_2. Fill it with dry argon.

2. Purify pyrrole by passing 0.5 ml pyrrole (as received) through a neutral Al_2O_3 column to remove any coloured components.

3. Dissolve 56 µl pyrrole in 1.944 ml of 100 mM KCl rigorously excluding any O_2 (final concentration of pyrrole is 100 mM).

4. Add 1 mg/ml of a biological recognition element and transfer 2 ml of this solution into the electrochemical cell (continuous outflow of Ar).

5. Connect the first electrode of the array to the potentiostat and apply a potentiostatic pulse profile to the working electrode with potentials of +875 mV for 1 sec (deposition phase) and 0 mV for 10 sec (resting phase).

6. Disconnect the electrode from the potentiostat and connect the second electrode of the array to it. Remove the deposition solution and wash the electrode array with 100 mM KCl to remove enzyme that is only adsorbed.

7. Dissolve 56 µl pyrrole in 1.944 ml of 100 mM KCl rigorously excluding any O_2 (final concentration of pyrrole is 100 mM).

8. Add 1 mg/ml of a second biological recognition element and transfer 2 ml of this solution into the electrochemical cell (continuous outflow of Ar).

9. Apply a potentiostatic pulse profile to the working electrode with

---

potentials of +875 mV for 1 sec (deposition phase) and 0 mV for 10 sec (resting phase).

10. Disconnect the electrode from the potentiostat and connect the next electrode of the array to it. Remove the deposition solution and wash the electrode array with 100 mM KCl to remove enzyme that is only adsorbed.

11. For subsequent electrodes continue again with step 6.

References

1. Bartlett, P.N. and Cooper, J.M. (1993). *J. Electroanal. Chem.*, **362**, 1.
2. Deshpande, M.V. and Amalnerkar, D.P. (1993). *Prog. Polym. Sci.*, **18**, 623.
3. Bartlett, P.N. and Birkin, P.R. (1993). *Synth. Methods*, **61**, 15.
4. Trojanowicz, M. and Krawczyk, T.K.V. (1995). *Mikrochim. Acta*, **121**, 167.
5. Schuhmann, W. (1995). *Mikrochim. Acta*, **121**, 1.
6. Bilger, R. and Heinze, J. (1993). *Synth. Methods*, **55**, 1424.
7. Schuhmann, W., Strike, D., and Koudelka-Hep, M. Unpublished results.
8. Kranz, C., Ludwig, M., Gaub, H.E., and Schuhmann, W. (1995). *Adv. Mater.*, **7**, 38.
9. Schuhmann, W., Kranz, C., Wohlschläger, H., and Strohmeier, J. (1997). *Biosens. Bioelectron.*, **12**, 1157.
10. Umana, M. and Waller, J. (1986). *Anal. Chem.*, **58**, 2979.
11. Foulds, N.C. and Lowe, C.R. (1986). *J. Chem. Soc. Faraday Trans.*, **82**, 1259.
12. Fortier, G., Brassard, E., and Bélanger, D. (1988). *Biotechnol. Tech.*, **2**, 177.
13. Fortier, G., Brassard, E., and Bélanger, D. (1990). *Biosens. Bioelectron.*, **5**, 473.
14. Bartlett, P.N. and Whitaker, R.G. (1987). *J. Electroanal. Chem.*, **224**, 37.
15. Schuhmann, W. (1994). In *Diagnostic biosensor polymers* (ed. A.M. Usmani and N. Akmal). *ACS Symposium Series*, Vol. 556, p. 110.
16. Evans, S.E., Yon Hin, B.F.Y., and Lowe, C.R. (1992). *Proc. Cleveland-Symposium*, 115.
17. Wolowacz, S.E., Yon Hin, B.F.Y., and Lowe, C.R. (1992). *Anal. Chem.*, **64**, 1541.
18. Yon Hin, B.F.Y., Smolander, M., Crompton, T., and Lowe, C.R. (1993). *Anal. Chem.*, **65**, 2067.
19. Schuhmann, W. (1993). In *Proceeedings of Bioelectroanalysis, 2* (ed. E. Pungor), 11.10.–15.10.1992. Matrafüred, Ungarn. Akadémiai Kiadó, Budapest, 113.
20. Stefan, K.-P., Schuhmann, W., Parlar, H., and Korte, F. (1989). *Chem. Ber.*, **122**, 169.
21. Delabouglise, D. and Garnier, F. (1991). *N. J. Chem.*, **15**, 233.
22. Ho-Hoang, A., Fache, F., and Lemaire, M. (1996). *Synth. Commun.*, **26**, 1289.
23. Schuhmann, W., Wohlschläger, H., Kulys, J., and Schmidt, H.-L. Unpublished results.
24. Yon Hin, B.F.Y. and Lowe, C.R. (1994). *J. Electroanal. Chem.*, **374**, 167.
25. Schuhmann, W., Lammert, R., Uhe, B., and Schmidt, H.-L. (1990). *Sensors Actuators*, **B1**, 537.
26. Schuhmann, W. (1991). *Synth. Methods*, **414**, 29.
27. Schuhmann, W. and Schmidt, H.-L. (1992). In *Advances in biosensors*, Vol. II (ed. A.P.F. Turner), p. 79. JAI Press, London.

28. Schalkhammer, T., Mann-Buxbaum, E., Pittner, F., and Urban, G. (1991). *Sensors Actuators*, **B4**, 273.
29. Schalkhammer, T., Mann-Buxbaum, E., Urban, G., and Pittner, F. (1990). *J. Chromatogr.*, **510**, 355.
30. Schuhmann, W., Huber, J., Mirlach, A., and Daub, J. (1993). *Adv. Mater.*, **5**, 124.
31. Röckel, H., Huber, J., Gleiter, R., and Schuhmann, W. (1994). *Adv. Mater.*, **6**, 568.
32. Hiller, M., Kranz, C., Huber, J., Bäuerle, P., and Schuhmann, W. (1996). *Adv. Mater.*, **8**, 219.
33. Wang, J., Chen, S.-P., and Lin, M.S. (1989). *J. Electroanal. Chem.*, **273**, 231.
34. Cosnier, S., Deronzier, A., and Roland, J.F. (1991). *J. Electroanal. Chem.*, **310**, 71.
35. Yon Hin, B.F.Y., Sethi; R.S., and Lowe, C.R. (1990). *Sensors Actuators*, **B1**, 550.

$\boxed{\text{A1}}$

List of suppliers

Aldrich Chemical Company
Aldrich Chemical Company, PO Box 2060, Milwaukee, WI, USA.
Aldrich Chemical Company, The Old Brickyard, New Road, Gillingham, Dorset SP8 4XT, UK.
Amersham
Amersham International plc, Lincoln Place, Green End, Aylesbury, Buckinghamshire HP20 2TP, UK.
Amersham Corporation, 2636 South Clearbrook Drive, Arlington Heights, IL 60005, USA.
Anderman
Anderman and Co. Ltd., 145 London Road, Kingston-Upon-Thames, Surrey KT17 7NH, UK.
J. T. Baker, 222 Red School Lane, Phillipsburg, NJ 08865, USA.
Beckman Instruments
Beckman Instruments UK Ltd., Progress Road, Sands Industrial Estate, High Wycombe, Buckinghamshire HP12 4JL, UK.
Beckman Instruments Inc., PO Box 3100, 2500 Harbor Boulevard, Fullerton, CA 92634, USA.
Becton Dickinson
Becton Dickinson and Co., Between Towns Road, Cowley, Oxford OX4 3LY, UK.
Becton Dickinson and Co., 2 Bridgewater Lane, Lincoln Park, NJ 07035, USA.
Belovo Chemicals, Industrial Area 1, 6600 Bastogne, Belgium.
Bio 101 Inc.
Bio 101 Inc., c/o Statech Scientific Ltd., 61–63 Dudley Street, Luton, Bedfordshire LU2 0HP, UK.
Bio 101 Inc., PO Box 2284, La Jolla, CA 92038–2284, USA.
Bio-Rad Laboratories
Bio-Rad Laboratories Ltd., Bio-Rad House, Maylands Avenue, Hemel Hempstead HP2 7TD, UK.
Bio-Rad Laboratories, Division Headquarters, 3300 Regatta Boulevard, Richmond, CA 94804, USA.
Boehringer Mannheim
Boehringer Mannheim UK (Diagnostics and Biochemicals) Ltd., Bell Lane, Lewes, East Sussex BN17 1LG, UK.

List of suppliers

Boehringer Mannheim Corporation, Biochemical Products, 9115 Hague Road, PO Box 504, Indianopolis, IN 46250–0414, USA.

Boehringer Mannheim Biochemica, GmbH, Sandhofer Str. 116, Postfach 310120, D-6800 Ma 31, Germany.

British Drug Houses (BDH) Ltd., Poole, Dorset, UK.

Calbiochem Novabiochem AG, Weidenmattweg 4, Postfach CH-448 Läufelfingen, Switzerland.

Difco Laboratories

Difco Laboratories Ltd., PO Box 14B, Central Avenue, West Molesey, Surrey KT8 2SE, UK.

Difco Laboratories, PO Box 331058, Detroit, MI 48232–7058, USA.

Dow Chemical, 2020-T Willard H. Dow Center, Midland, MI 48674, USA.

Du Pont

Dupont (UK) Ltd., Industrial Products Division, Wedgwood Way, Stevenage, Hertfordshire SG1 4Q, UK.

Du Pont Co. (Biotechnology Systems Division), PO Box 80024, Wilmington, DE 19880–002, USA.

European Collection of Animal Cell Culture, Division of Biologics, PHLS Centre for Applied Microbiology and Research, Porton Down, Salisbury, Wiltshire SP4 0JG, UK.

Falcon (Falcon is a registered trademark of Becton Dickinson and Co.)

Fisher Scientific Co., 711 Forbest Avenue, Pittsburgh, PA 15219–4785, USA.

Flow Laboratories, Woodcock Hill, Harefield Road, Rickmansworth, Hertfordshire WD3 1PQ, UK.

Fluka

Fluka Chemie AG, Industriestrasse 25, CH-9471, Buchs, Switzerland.

Fluka Chemicals Ltd., The Old Brickyard, New Road, Gillingham, Dorset SP8 4JL, UK.

Fluka Chemical Corp., 980 South 2nd Street, Ronkonkoma, NY 11779–7238, USA.

Fujikura Kasei Company, Ltd., 2–6–15 Shibakoen, Minato-ku, Tokyo 105, Japan.

Gibco BRL

Gibco BRL (Life Technologies Ltd.), Trident House, Renfrew Road, Paisley PA3 4EF, UK.

Gibco BRL (Life Technologies Inc.), 3175 Staler Road, Grand Island, NY 14072–0068, USA.

Arnold R. Horwell, 73 Maygrove Road, West Hampstead, London NW6 2BP, UK.

Hybaid

Hybaid Ltd., 111–113 Waldegrave Road, Teddington, Middlesex TW11 8LL, UK.

Hybaid, National Labnet Corporation, PO Box 841, Woodbridge, NJ 07095, USA.

HyClone Laboratories, 1725 South HyClone Road, Logan, UT 84321, USA.

International Biotechnologies Inc., 25 Science Park, New Haven, Connecticut 06535, USA.

Invitrogen Corporation

Invitrogen Corporation, 3985 B Sorrenton Valley Building, San Diego, CA 92121, USA.

Invitrogen Corporation, c/o British Biotechnology Products Ltd., 4–10 The Quadrant, Barton Lane, Abingdon OX14 3YS, UK.

Kodak: Eastman Fine Chemicals, 343 State Street, Rochester, NY, USA.

Life Technologies Inc., 8451 Helgerman Court, Gaithersburg, MN 20877, USA.

Merck

Merck Industries Inc., 5 Skyline Drive, Nawthorne, NY 10532, USA.

Merck, Frankfurter Strasse, 250, Postfach 4119, D-64293, Germany.

Millipore

Millipore (UK) Ltd., The Boulevard, Blackmoor Lane, Watford, Hertfordshire WD1 8YW, UK.

Millipore Corp./Biosearch, PO Box 255, 80 Ashby Road, Bedford, MA 01730, USA.

Molecular Probes

Molecular Probes, Inc., PO Box 22010, Eugene, OR 97402–0469, USA.

Molecular Probes Europe BV, PoortGebouw, Rijnsburgerweg 10, 2333 AA Leiden, The Netherlands.

Morton Thiokol, Morton International, 100 North Riverside Plaza, Chicago, IL 60606–1596, USA.

New England Biolabs (NBL)

New England Biolabs (NBL), 32 Tozer Road, Beverley, MA 01915–5510, USA.

New England Biolabs (NBL), c/o CP Labs Ltd., PO Box 22, Bishops Stortford, Hertfordshire CM23 3DH, UK.

Nikon Corporation, Fuji Building, 2–3 Marunouchi 3-chome, Chiyoda-ku, Tokyo, Japan.

Nunc, 2000 North Aurora Road, Naperville, IL 60563–1796, USA.

Perkin-Elmer

Perkin-Elmer Ltd., Maxwell Road, Beaconsfield, Buckinghamshire HP9 1QA, UK.

Perkin Elmer Ltd., Post Office Lane, Beaconsfield, Buckinghamshire HP9 1QA, UK.

Perkin Elmer-Cetus (The Perkin-Elmer Corporation), 761 Main Avenue, Norwalk, CT 0689, USA.

Pharmacia Biotech Europe, Procordia EuroCentre, Rue de la Fuse-e 62, B-1130 Brussels, Belgium.

Pharmacia Biosystems

Pharmacia Biosystems Ltd. (Biotechnology Division), Davy Avenue, Knowlhill, Milton Keynes MK5 8PH, UK.

Pharmacia LKB Biotechnology AB, Björngatan 30, S-75182 Uppsala, Sweden.

Pierce
Pierce Chemical Co., 3747 North Meridian Road, PO Box 117, Rockford, IL 61105, USA.
Pierce Europe BV, PO Box 1512, 3260 BA-Oud Beijerland, The Netherlands.
Pierce and Warriner, 44 Upper Northgate Street, Chester CH1 4EF, UK.
Polysciences, Inc., 400 Valley Road, Warrington, PA 18976, USA.
Promega
Promega Ltd., Delta House, Enterprise Road, Chilworth Research Centre, Southampton, UK.
Promega Corporation, 2800 Woods Hollow Road, Madison, WI 53711–5399, USA.
Qiagen
Qiagen Inc., c/o Hybaid, 111–113 Waldegrave Road, Teddington, Middlesex TW11 8LL, UK.
Qiagen Inc., 9259 Eton Avenue, Chatsworth, CA 91311, USA.
Rainin Instrument Co. Inc., 5400 Hollis Street, Emeryville, CA 94608–2508, USA.
Schleicher and Schuell
Schleicher and Schuell Inc., Keene, NH 03431A, USA.
Schleicher and Schuell Inc., D-3354 Dassel, Germany. Schleicher and Schuell Inc., c/o Andermann and Company Ltd.
Shandon Scientific Ltd., Chadwick Road, Astmoor, Runcorn, Cheshire WA7 1PR, UK.
Shipley Chemicals, Herald Way, Coventry CV3 2RQ, UK.
Sigma Chemical Company
Sigma Chemical Company (UK), Fancy Road, Poole, Dorset BH17 7NH, UK.
Sigma Chemical Company, 3050 Spruce Street, PO Box 14508, St. Louis, MO 63178–9916, USA.
Sorvall DuPont Company, Biotechnology Division, PO Box 80022, Wilmington, DE 19880–0022, USA.
SPA (Società Prodotti Antibiotici S.p.A.),Via Biella 8, 20143 Milano, Italy.
STC Laboratories Inc., 348 Saulteaux Crescent, Winnipeg, MA R3J 3T2, Canada.
Stratagene
Stratagene Ltd., Unit 140, Cambridge Innovation Centre, Milton Road, Cambridge CB4 4FG, UK.
Stratagene Inc., 11011 North Torrey Pines Road, La Jolla, CA 92037, USA.
United States Biochemical, PO Box 22400, Cleveland, OH 44122, USA.
Vector Laboratories
Vector Laboratories Ltd., 16 Wulfric Square, Bretton, Peterborough PE3 8RF, UK.
Vector Laboratories, Inc., 30 Ingold Road, Burlingame, CA 94010, USA.
Wellcome Reagents, Langley Court, Beckenham, Kent BR3 3BS, UK.

Index